INTERNATIONAL CENTRE FOR MECHANICAL SCIENCES

COURSES AND LECTURES · No. 276

NOISE GENERATION AND CONTROL

IN MECHANICAL ENGINEERING

P.O.A.L. DAVIES
UNIVERSITY OF SOUTHAMPTON

M. HECKL
TECHNICAL UNIVERSITY OF BERLIN

G.L. KOOPMAN
UNIVERSITY OF HOUSTON, TEXAS

EDITED BY

G. BIANCHI
POLITECNICO DI MILANO

SPRINGER-VERLAG WIEN GMBH

Per la stampa di questo volume il CNR ha assegnato il contributo n. 206072/07/67078.

ISBN 978-3-211-81710-0   ISBN 978-3-7091-2894-7 (eBook)
DOI 10.1007/978-3-7091-2894-7

FOREWORD

The acknowledged importance of the dangers of noise pollution demands that the engineer, and in particular the mechanical engineer, be able not only, and not so much, to find remedies to the problem as it presents itself in each case, but that he be, above all, "sensitive" to the problem of noise in all stages of machine design and installation. This "sensitivity" can give useful fruits only if it is based upon an intimate knowledge of the phenomena of noise production and propagation in machine and plants installation.

This book is a collection of lectures given on these themes in one of the advanced courses we have organized at the International Centre for Mechanical Sciences.

A view of the basic definitions of acoustic quantities and the techniques for their representation and measurement is given by G.L. Koopman, who then presents the main body of pertinent acoustic theory: wave acoustics, ray acoustics, effects of boundaries, acoustic sources. The theory is then applied to a typical, conventional source of noise, the motor vehicle, which is thoroughly analyzed with advanced techniques for the study of body interaction.

P.O.A.L. Davies treats the question of flow associated noise, with applications to common problems in industrial engineering, such as valves, fans, blowers, compressors. Particular attention is given to the problems of sound generation and propagation in flow ducts and to that of aircraft noise.

M. Heckl considers the problem of the transmission and propagation of sound in structures. The effects of form and materials on the transmission of vibratory energy and the possibilities of damping are studied. The mechanisms of the radiation of sound from structures, the loss of transmission in walls, the possibilities of damping sound with opportunely selected materials are also examined.

The volume, for its very origin as a collective lecture series, cannot presume to give an exhaustive and homogeneous treatment of all noise problems in mechanical engineering. It does indicate a correct approach to the rigorous examination of fundamental problems as experienced by active researchers in the field.

Giovanni Bianchi
Secretary General of CISM

CONTENTS

SOUND, VIBRATION AND SHOCK I

Measuring Systems

When converting the wave motion of sound and vibration into a measurable quantity for the purpose of analysis, it is necessary to use combinations of several instruments which together make up the overall measurement system. A typical layout of possible combinations is shown in figure 1.

1. Sensing Transducer and Signal Conditioner

The transducer, as the name implies, is a device which converts one form of energy into another. In sound and vibration analysis, sensing transducers are used for the conversion of acoustic or vibratory motion into electrical energy, the most convenient form for measurement, recording, and analysis. Basically, sensing transducers fall into the following two electrical classifications:

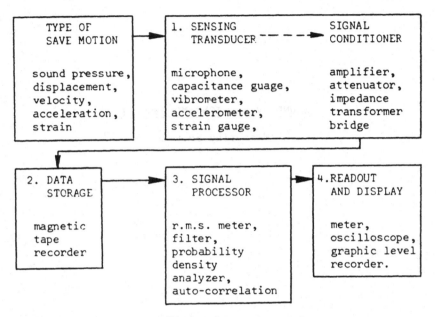

Figure 1.

(a) Non-generating devices which require a supply voltage. The
     mechanical response of this device causes a change in capaci-
     tance, inductance, or resistance in its electrical network and
     an output signal is produced.

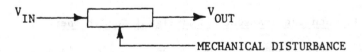

(b) Generating devices which do not require a separate supply
     voltage. These devices contain an element which generates a
     voltage proportional to the relative velocity or a charge
     proportional to the strain in the element.

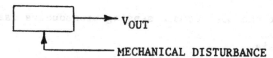

The transfer function which relates the input mechanical disturbance to the output voltage of the transducer is always frequency and
amplitude dependent: The three main characteristics which describe the
operation of a transducer are thus (1) sensitivity, (2) dynamic range,
and (3) frequency response. These are illustrated in Figure 2.

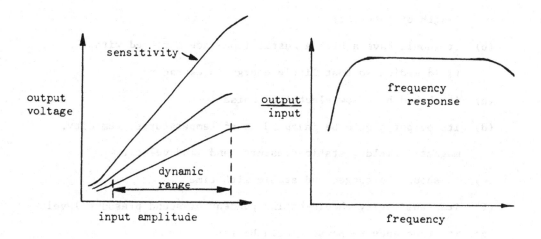

Figure 2

To achieve a practical range of amplitudes and frequencies over which a
transducer can operate, it is necessary to couple it electrically to a
signal conditioner which is used to amplify, attenuate or transform the
generated output voltage. Thus, in specifying the transfer function of
a transducer, it is usual to include the effect of the signal conditioner
in describing the overall response characteristics.

With the above brief introduction, let us next examine in detail a
few of the more common sensing transducers used in sound and vibration
analysis.

1.1 Pressure Sensing Transducers

Generally, an ideal microphone should have the following characteristics:

(a) it should cause negligible diffraction of the sound field (i.e.
     its dimensions should be small compared with the smallest wave-
     length of interest).

(b) it should have a high acoustic impedance compared with the
     fluid medium so that little energy is extracted.

(c) it should have low electrical noise

(d) its output should be independent of temperatures, humidity,
     magnetic fields, static pressure, and wind velocity

(e) it should be rugged and stable with time

(f) its sensitivity should be independent of sound pressure level

(g) its frequency response should be flat

(h) it should introduce a zero phase shift between the sound
     pressure and the electrical output signal.

Pressure sensitive microphones can be divided into three basic types:
Condenser (strictly speaking capacitor), piezo-electric (or ceramic),
and moving coil (or dynamic).

a) Condenser Microphone

Method of operation (see figure 3) - a stretched thin metal diaphragm moves under sound pressure variation relative to a fixed

insulated plate. The variation in capacitance between them causes
a signal voltage proportional to the sound pressure to appear
across the load resistance when the diaphragm and plate are polarized from a constant voltage source (usually of the order of 200
volts).

<u>Advantages</u> - uniform sensitivity

            - low distortion

            - excellent frequency response

<u>Disadvantages</u> - low sensitivity (no longer a problem with advent of
            large low distortion amplifiers

            - high internal impedance

            - requires polarization voltage

            - sensitive to conditions of high humidity.

<u>Applications</u> - precise acoustic measurements

            - high fidelity sound recording

<u>Example of Specifications</u>

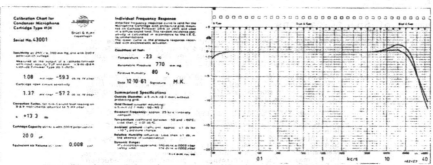

Fig. 3 Reproduction of the calibration chart supplied with a Condenser Microphone Cartridge.

b) Piezo-electric Microphone

Method of operation (see Figure 4) - the piezo-electric micro-phone uses a ceramic (lead titanate, lead zirconate) as the voltage-generating element. A diaphragm fastened to the ceramic transfers the sound-pressure variations into a corresponding varying force that deforms the ceramic element. The type of deformation may be shear, compression, or bending, arranged by coupling the microphone diaphragm to an end or corner of the element.

Advantages - high stability, low sensitivity to high temperatures

or high humidity.

- rugged and inexpensive.

- low internal impedance

- low cost.

Disadvantages - restricted upper frequency response

- production of spurious response at high acceleration

levels.

Applications - industrial sound level meters.

Example of Specifications

Frequency: Designed for flat random-incidence response (see curves). Typically flat, $\pm$ 1 dB, from 5 Hz to 20 Hz re the 500-Hz level. Time constant of pressure-equalizing leak is typically 0.08 s with a corresponding 3-dB roll-off at 2 Hz.

Sensitivity: −60 dB nominal, −62 dB min, re 1 V/μbar. Temperature coefficient, ≈ −0.01 dB/°C. Maximum sound-pressure level, at 150 dB SPL distortion is <1%; at >+ 174 and >−184 dB SPL peak, microphone fails.

Impedance: 1560-P5, 385 pF $\pm$ 15% at 23°C; 1560-P6, 405 pF $\pm$ 15% at 23°C. Temperature coefficient of both, 2.2 pF/°C from 0 to 50°C.

Environmental: No damage from −55 to +60°C, 0 to 100% RH; at 95°C, a 0.5-dB permanent sensitivity loss may occur.

Mechanical: Terminals, 3-terminal microphone connector; both terminals may be floated with respect to ground for hum reduction. Dimensions, cartridge only, 1⅛ in. (28.6 mm) long, 0.936 $\pm$ 0.002 in. (23.7 mm $\pm$ 50 um) dia; 1560-P5 assembly, 2⅛ (58.7 mm) long, ¹⁵⁄₁₆ in. (23.7 mm) dia; 1560-P6 assembly, 11¾ in: (298 mm) long, ¹⁵⁄₁₆ in. (23.7 mm) dia. Weight, 1560-P5, 2 oz (60 g) net, 1 lb (500 g) shipping; 1560-P6, 10 oz (300 g) net, 2 lb (900 g) shipping.

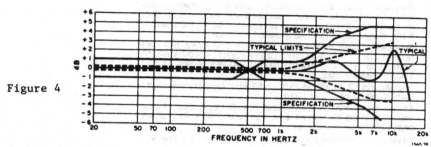

Figure 4

c) Electret Microphone

Method of operation (see Figure 4a) - similar in operation to the capacitor microphone. However, the polarization voltage is provided by a permanently polarized electret foil diaphragm. The electret diaphragm consists of a stretched plastic foil with a thin metal coating (usually gold) on the outer surface.

Advantages - no polarization voltage needed

      - rugged construction

      - large capacitance

      - low cost

Disadvantages   less smooth frequency response at high frequencies

      - upper temperature limit of about $50^{\circ}C$

      - polarization may not be sufficiently stable over long periods of time (several years).

Applications communication systems

Example of Specification

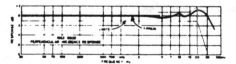

Figure 4a

Frequency: Curve shows typical response and guaranteed limits; individual response curve supplied with each microphone. Below 20 Hz, the microphone is typically flat ±1 dB down to 5 Hz. Response is essentially omnidirectional.
Sensitivity Level: NOMINAL: −56 dB re 1 V/N/m² (−76 dB re 1 V/μbar). TEMPERATURE COEFFICIENT: ~ +0.03 dB/°C from 0 to +55°C. MAXIMUM SOUND-PRESSURE LEVEL: 170 dB absolute max.
Impedance: 12 ±1 pF, at 25°C and 1 kHz; temperature coefficient < +0.02 pF/°C at 1 kHz.
Environment: −20 to +55°C and 90% RH operating; 1-year exposure in an environment of +55°C and 90% RH causes negligible sensitivity change.
Vibration Sensitivity: 83 dB equivalent SPL from 1 g (perpendicular to diaphragm) at 20 and 100 Hz.

1.2   Mechanical Vibration (and Force) Sensing Transducers

Vibration sensing transducers can be grouped generally into two categories, namely, contacting and non-contacting. The choice of which type to use depends entirely on the requirements of the particular experiment.

While the use of the contacting transducer does eliminate the need for (and the problem of) designing a vibration isolated mounting fixture as in the case of the non-contacting transducer, the direct contact of the transducer may influence the overall response of the vibrating system. This is especially true when the mass of the transducer is appreciable compared with that of the system. An ideal transducer would have the following characteristics:

(a)   ease of mounting (is a fixed reference frame required?)

(b)   no contribution to the dynamics of the vibrating system being investigated

(c)   durability and ruggedness.

(d)   flat frequency response

(e) low electrical noise

(f) its sensitivity should be independent of vibration amplitude.

(g) its output should be independent of temperature, humidity,
     and magnetic fields

(h) it should introduce a zero phase shift between the sound pressure
     and the electrical output signal.

Let us next examine a few of the more common transducers used in
vibration analysis.

a) Contacting Seismic Transducer

(i) Accelerometer, (see Figure 5) - this device consists of a mass
     supported in the sensitive direction by a strain-sensitive
     element, frequently a piezo-electric ceramic. The case, in
     which the mass is thus seismically suspended, is rigidly
     attached to the vibrating body. The inertia of the mass
     causes it to experience a force proportional to the acceleration
     of the vibrating body provided the vibration frequency is well
     below the resonant frequency of the mass or the slightly
     flexible 'ceramic' spring. The voltage generated is thus
     proportional to the acceleration of the vibrating body.
     Accelerometers have the advantages of small size, light weight,
     and wide frequency range.

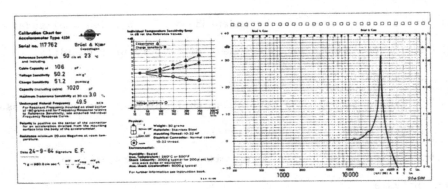

Fig. 5   Typical calibration chart as supplied with each accelerometer.

(ii) Vibrometers – although in principle of construction similar to accelerometers, these devices measure in a frequency range above the spring-mass resonant frequency of the system. The effect of this is that the mass remains virtually stationary (seismic) and the transducer measures the relative velocity or displacement between the seismic mass and vibrating frame of the instrument.

b) Non-contacting Transducers

(i) Velocity Gauges (see figure 6) – this is a variable reluctance device which consists of a cylindrical permanent magnet on which is wound an insulated low resistance wire. When placed a few millimeters from a metallic test piece, such a device will produce a signal which is proportional to the separation distance. This device can also be used reciprocally as a vibration exciter.

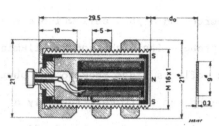

Fig. 6. Section of the Magnetic Transducer with in-
dication of the main dimensions in mm (1 mm = 0.0394").
One high permeability disc is represented.

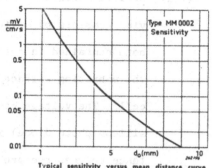

Typical sensitivity versus mean distance curve,
measured at 50 c/s with a constant RMS speed of
50 cm/sec.

(ii)   Displacement Gauge (see Figure 7) - this displacement sen-

sitive transducer consists of a shielded electrode which is

mounted parallel to the test piece and at a suitable distance

from it. The test piece forms the other electrode of the air-

gap capacitor. A polarization voltage (usually several hun-

dred volts) is applied across the electrodes. As the test

piece moves, an alternating voltage is produced which is

proportional to the displacement. One advantage of this trans-

ducer is that it can be calibrated statically by mounting it on a

traversing mechanism and measuring the voltage produced while the

displacement change is measured with a micrometer screw gauge.

Figure 7

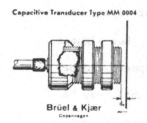

Capacitive Transducer Type MM 0004

Brüel & Kjær
Copenhagen

SPECIFICATIONS

Sensitivity. Inversely proportional to the distance (d_o) between the electrode of the MM 0004 and the test specimen. At $d_o = 0.5$ mm and with a displacement of 10^{-3} cm (peak to peak) the output voltage will be about 0.9 volt RMS.

Frequency Range: 20–200000 Hz.

Impedance: inversely proportional to the distance, d_o, between electrodes. At $d_o = 0.5$ mm the capacity is approximately 1.5 pF.

Static force, between transducer and test specimen. Approximately 56 mg at $d_o = 0.5$ mm. Inversely proportional to the square of the distance d_o.

Distortion: Increases inversely with the distance, d_o, between electrodes. Therefore d_o should be more than ten times the displacement (peak) and not less than 0.5 mm to keep the distortion below 5 %.

Polarization voltage: 200 V.

Electrode Area: 0.78 cm² (0.125 sq.inch).

Diameter (total): 21 mm (0.84 inch).

Length (total): 34 mm (1.36 inch).

Weight (excl. cable): 46.9 g (1.66 oz.)

(iii)   Optical Vibration Sensor - this displacement measuring de-
        vice consists of a light source and a photocell which are
        both contained in the same probe. The light source illuminates
        the vibrating surface which reflects light back to the photo-
        cell. The voltage generated by the photo cell (usually a
        silicon photo diode) is directly proportional to the dis-
        tance between the probe and the reflecting surface. Since
        the photocell has an extremely short rise time, the upper
        limiting frequency of this device is quite high. Also, it may
        be used on insulating and conducting materials and very little
        surface preparation is required.

2. Data Storage - Magnetic Tape Recorder

The magnetic recording of analogue or digital signals has the following attractions:

a)   the recorded and reproduced signals are electrical in form and thus
the recorder may be inserted directly into the measurement system.

b)   a wide range of bandwidths and recording time is available.

c)   simultaneous recording or reproduction of more than one signal may
be achieved.

d)   time-base (on bandwidth) compression or expansion of signal is
possible by changing the magnetic tape speed on replay. Thus signals
with bandwidths outside that of the signal analysis equipment can be
brought into the equipment bandwidth. Alternatively, signals may be
"speeded up" on replay to reduce analysis time or reproduced slowly for

examination of signal in detail by a pen recorder for example.

e) signals can be stored for further examination at some later time.
This is frequently done in field measurements using a portable recorder
to acquire the signal which is subsequently analyzed in the laboratory
elsewhere.

Although digital signal processing is taking over some of the above
roles, the magnetic recorder is frequently used especially in 'e'.
Some understanding of the magnetic recorder is essential if excessive
signal error is to be avoided in the record/reproduce process. The
recording process is complex and some simplifications will be made in
these notes.

Firstly, it should be realized that the recorder is the "weak
link" in the signal measurement chain and therefore direct on-line measurement without recording should be made whenever possible - a recorder being used perhaps only as a back-up.

In cases where use of the recorder is essential, the following areas
must be understood and procedures adopted to keep signal measurement
error to an acceptable level.

(a) Matching input signal to recorder input

(b) Monitoring input signal amplitude.

(c) Matching recorder output to external equipment.

(d) Recorder bandwidth, phase-shift; wave form distortion; non-
     linearity, signal to noise ratio/dynamic range; time/phase
     shift between signal channels; and cross-talk.

(e) Time-base errors resulting from errors in tape speed.

(f) Compatibility with other recorders - additional errors to (d)

   and (e).

(g) Calibration procedures.

As the above are interactive, it is useful to start with the recorder considered in two parts, namely, the tape transport mechanism,
and the electronic recording and replay system.

a) Tape Transport System

   Reliability and overall performance requires good precision engineering and adequate maintenance of the tape transport system which
drives the magnetic tape over the recording and replay heads at as near
to constant speed as possible. Rewinding facilities and braking systems

are also provided. Facilities for battery (portable) operation and a
range of tape speeds may also be provided.

   Tape recorders may be broadly classified by the following
   groupings:-

Instrumentation - Wide range of bandwidths and tape speeds available
   with good control of tape speed.

Professional Audio - Bandwidth limited to audio frequency range (30Hz -
   30kHz); limited range of tape speeds; linearity and signal to
   noise ratio is optimized and can give better noise levels than the
   instrumentation type. Good control of tape speed.

Domestic - similar to group (ii) but of variable performance approaching
   professional at best but ranging to very poor. For this and other

reasons such lack of standardization or poor speed control, these
recorders have limited application and required careful selection
for noise/vibration instrumentation applications.

b) Electronic Recording and Replay System

Two different recording systems are commonly used in sound and
vibration measurements, namely, 'direct recording' (D.R) and 'frequency
modulation' (F.M). D.R permits a lower tape speed (a quarter or less)
to be used for a given upper frequency response and hence gives more
recording time per spool of tape. However the low frequency response
does not extend below about 20Hz and if a low response is required or
if the waveform of the recorded signal must be preserved (e.g) sonic-
boom, a F.M. system is necessary. With F.M the reproduced signal is
mainly dependent on the electronic circuitry. Frequency response and
gain calibration is easily better than $\pm$ 0.5 dB and recordings may be
reproduced on other machines of similar specification but with smaller
machines the low tape speeds available may limit the bandwidth to 2.5 or
5.0 kHz. Accurate matching of the recorder and reproducer recording
characteristics is essential for good results with D.R and so recording
and reproduction on the same machine is highly desirable.

The overall recording characteristic is influenced by the type of
recording tape, replay head manufacturing tolerance, mechanical align-
ment of the head, the high frequency bias level during recording, wear
and dirt build-up at the head surface, tape curl and the form of recording frequency equalisation employed. The bandwidth of all recorders
increases approximately linearly with tape speed.

Figures 8 and 9 show the typical frequency response curves of a direct recording and a frequency modulation system.

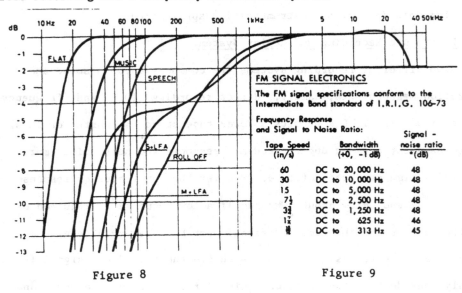

| Figure 8 | Figure 9 |

3. Signal Processing

The term 'signal processing' covers a vast range of possible operations depending upon the nature of the signal and the information for which it is being processed or analyzed. For a large majority of sound and vibration signals, it is sufficient to measure the r.m.s. level and the frequency spectrum. The following equipment is used to measure these quantities:

3.1 Sound Level Meter

The basis instrument of a sound measuring system is the sound-level meter. It is a portable meter for reading the sound level at its

microphone in terms of a standard reference pressure (20μPa). This is
an overall level and provides no information on spectral content.
Basically, the instrument consists of an omni-directional microphone, a
calibrated attenuator, a stabilised amplifier, an indicating meter, and
weighting networks. The networks provide the three common sound-level
meter responses (A, B, and C).

3.2 Analysers

A number of analysers are available for use with the sound level
meter or for use with microphones and vibration transducers directly or
with preamplifiers. These analysers vary in cost, complexity, and ease
of operation. Choice among them is generally determined by the amount of
detailed information needed, the speed of processing required, the nature
of the output format, and the auxiliary processing that may be required.

The analysis of a signal into its distribution across the frequency
spectrum can be performed in very fine or broad frequency bands,
according to the characteristics of the source and the use to which the
data will be put. The first level of information on spectral content is
obtained from analyzing the signal in octave bands. Octave bands have
constant percentage bandwidths with the upper frequency limit of each
band twice the value of its lower frequency limit. The audible range of
frequencies is covered in ten octaves.

The next level of detail is based on one-third octave bands, where
one-third octave is the interval between two frequencies with the ratio
cube root of two. The current standard for the spectrum allocation into

octave and third-octave bands is the preferred series given in Table 1.

Preferred frequencies	One oct.	Half oct.	Third oct.	Preferred frequencies	One oct.	Half oct.	Third oct.	Preferred frequencies	One oct.	Half oct.	Third oct.
16	X	X	X	160			X	1,600			X
18				180		X		1,800			
20			X	200			X	2,000	X	X	X
22,4		X		224				2,240			
25			X	250	X	X	X	2,500			X
28				280				2,800		X	
31.5	X	X	X	315			X	3,150			X
35.5				355		X		3,550			
40			X	400			X	4,000	X	X	X
45		X		450				4,500			
50			X	500	X	X	X	5,000			X
56				560				5,600		X	
63	X	X	X	630			X	6,300			X
71				710		X		7,100			
80			X	800			X	8,000	X	X	X
90		X		900				9,000			
100			X	1,000	X	X	X	10,000			X
112				1,120				11,200		X	
125	X	X	X	1,250			X	12,500			X
140				1,400		X		14,000			
160			X	1,600			X	16,000	X	X	X

Table 1 Preferred Center Frequencies for Noise Measurements

A series of octave bands widely used in the past ranged 75 – 150 Hz, 150 – 300 Hz, etc, and much published data still exists in this form.

The most detailed spectral distribution information is the spectrum level. Pressure spectrum level is the effective sound pressure level of the sound energy contained in a bandwidth of 1 Hz centered at a specified frequency; power spectrum level is the power level in a band 1 Hz wide centered at a specified frequency. Rarely is such detail

as from 1 Hz bandwidth analysis required. Between the two extremes of
octave band level and spectrum levels are numerous nonstandardized analysis
bands, of two types: (1) constant percentage bandwidths and (2) constant
bandwidths of equal frequency increments. A 6Hz bandwidth filter usually
gives more than adequate definition unless the data extends to unusually
low frequencies (< 50 Hz). The choice of level of detail for analysis
of s signal into frequency bands must be based on the characteristics of
the source. The two extremes of source spectral characteristics are
white noise (equal energy per unit frequency bandwidth over a specified
total frequency band), and pure tones (single-fréquency sine waves). In
interpreting acoustical data, it is necessary to bear in mind that,
depending on the source, the various degrees of resolution can yield
very different results for the same signal input. As an example,
figure 10 compares a real spectrum (for a continuous random signal with
a superposed sinusoidal signal at 400 Hz) with the apparent spectra
obtained for various degress of resolution.

3.3 Real Time Analysers

    Recent developments in analyser design have produced a completely
new concept of analyser called the Real Time Analyser. These analysers
can perform all the functions of the types mentioned previously, and
with the aid of a small computer several functions hitherto only
possible by the use of expensive computer complexes. These analysers
again come in two distinct forms: (a) third octave (constant percentage)
and (b) narrow band.

(a) One Third Octave Real Time Analysers work on similar lines to any

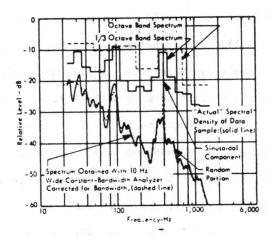

Figure 10   Comparison of the Outputs of Several
Levels of Spectral Analysis for the
Same Input Signal.

conventional one-third-octave filter, except that instead of switching

individual filters, each analyser has a complete set of filters covering

the audio frequency spectrum, or where fractional filters are employed,

selected parts of the range. The output from each filter is detected,

held in an electronic memory and displayed simultaneously, a complete

spectrum taking only about 1 ms (1/1000th of a second) to produce. The

memory can then be "unloaded" into various display devices at speeds

compatible with the device. The memory can also be continusously up-

dated with fresh information, or new information added to that already

existing to produce an averaged spectrum.

(b) Narrow Band or Time Compression Real Time Analysers work on some-

what different techniques. Here the desired audio range is stepped by

a constant bandwidth filter in "compressed time" to produce a narrow band

spectrum which again can be stored for unloading into many display devices. A typical configuration for such an analysis would be 0-2kHz in
500 steps each having a bandwidth of 4 Hz. As with the other type of
analyser, spectra can be averaged, and with additional electronics
synthesised into octaves or one-third-octaves.

Both the analysers described above can perform analyses in very
short periods of time, hence obviating the need for stable conditions,
or tape loops. They can be interfaced with a computer to perform very
complex tasks and can provide plots on a wide variety of displays.
They are, however, considerably more expensive and require more skilled
setting up.

4. Readout and Display

4.1 Meters

Vibration and sound level meters provide the simplest form of
readout and display. Generally, the meters can be selected to indicated
r.m.s at true peak-to-peak values of the incoming signal. R.M.S indica-
tion is employed for most measurements because of its relationship
with the energy context of the signal. Peak-to-peak is often used for
measuring shocks and for vibration measurement in the displacement mode.

4.2 Graphic Level Recorders

For permanent records of sound level measurements, particularly
where time dependence is of interest, sound level recorders find general
use. Sound reverberation decay and monitoring of traffic noise are
examples of such uses.

Recording is made on paper rolls by a stylus or pen driven by a moving coil in the field of a permanent-magnet, and arrangement similar to that of a loudspeaker voice coil. The stylus is also connected to a logarithmic potentiometer. At zero signal the stylus is held at the limit of its travel at which the potentiometer has zero attenuation. On the introduction of a signal, the amplifier produces a current in the moving coil which moves the stylus across the paper. At the same time, the attached slider on the input potentiometer moves in a direction to attenuate the input to the driver amplifier, so that a balance position for the slider and stylus is reached, proportional to the input signal in dB. A feedback coil and switched variable resistance provide an adjustment for maximum writing speed by varying the amount of velocity damping. This makes it possible to avoid excessive overshoot on sudden changes in signal level, or to damp out unwanted high frequencies in the recording. Most recorders can be used in conjunction with automatically switched filters to plot frequency or sound transmission loss spectra on specially calibrated paper. Frequency response curves can also be taken, for example for loudspeakers, using a compression circuit to keep power level to the speaker constant.

4.3 Oscilloscopes

Two advantages stand out in the use of an oscilloscope as a display device for sound signals: an extremely fast rise time (frequency response well above the audio-frequency range), and the ability to 'see' the signal being measured, and monitor changes directly.

Against this, recording of traces is somewhat cumbersome since it involves photography, although short-term recording for detailed observation is possible with storage oscilloscopes, which retain the image for a long enough time to allow it to be traced or photographed at leisure.

Combined with this storage facility, the cathode-ray oscilloscope is also used as a display for 'instantaneous' or 'real-time' spectrum analysers, which divide the audio-frequency spectrum into one-third octave channels. Repetitive scanning of each channel, which has its own detector continously connected to the input signal, gives band levels very shortly after the actual detection period and continuously changing as the input changes. A storage mode may be switched, to 'freeze' the display for inspection or photographing.

References

1. HARRIS, C.M. 1957 McGraw-Hill Book Company Handbook of Noise Control.

2. HARRIS, C.M. and CREDE, C.E. 1961 McGraw-Hill Book Company Shock and Vibration Handbook Vol. 1: Basic Theory and Measurements; Vol. 2: Data Analysis, Testing, and Methods of Control.

3. BERANEK, L.L. 1971 McGraw-Hill Book Company Noise and Vibration Control.

4. BROCH, J.T. 1971 Bruel & Kjaer Acoustic Noise Measurements.

5. CROCKER, M.J. and PRICE, A.J. 1975 CRC Press Inc Noise and Noise Control Volume I.

SOUND, VIBRATION AND SHOCK II

Spectra

In sound and vibration analysis, one encounters a wide variety of
wave forms ranging from non-deterministic signals produced by such
processes as combustion or turbulent flow to the more deterministic
ones observed in say, rotating machinery or electrical transformers.
For purposes of analysis, the actual time history of the signal in its
original, unprocessed form offers little information and it is necessary
to transform the time domain description of the signal into the frequency domain where the energy distribution of the signal can be
quantified as a function of frequency. Once this information is
available, it then becomes possible to identify which components
of a system act as sources of unwanted sound and/or vibration.

The transformation of deterministic wave forms from the time to frequency domain is accomplished by the method of Fourier which involves either series or integral relationships. For very short pulses, such as in shock phenomena, however, it is useful to represent the transient signal in the frequency domain in the form of shock spectra. This transformation is carried out with a form of the Duhamel integral.

For non-deterministic (random) wave forms, the frequency domain representation of a signal is obtained not by Fourier transforming the signal itself, but instead transforming an indirect time domain description of signal in the form of an auto-correlation function. The resulting frequency domain representation of the signal is referred to as the mean square spectral density.

The following table summarizes the above time/frequency domain relationships:

TIME DOMAIN	TRANSFORM PROCESS	FREQUENCY DOMAIN
DETERMINISTIC WAVE FORMS		
PERIODIC	FOURIER SERIES	LINE SPECTRA
APERIODIC	FOURIER INTEGRAL	CONTINUOUS SPECTRA
SHOCK PULSES	DUHAMEL INTEGRAL	(INITIAL) SHOCK SPECTRA
NON-DETERMINISTIC WAVE FORMS		
RANDOM	FOURIER INTEGRAL	MEAN SQUARE
(AUTO-CORRELATION)		SPECTRAL DENSITY

1.   DETERMINISTIC WAVE FORMS

1.1 Periodic Functions

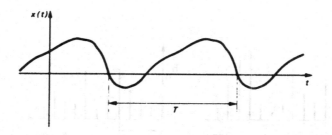

Figure 1 - arbitrary periodic function of time

Given that x(t) is a periodic function of time t with period T as shown in figure 1, x(t) can always be expressed in an infinite Fourier series of the form

$$x(t) = a_o + \sum_{n=1}^{\infty} (a_n \cos \frac{2\pi n t}{T} + b_n \sin \frac{2\pi n t}{T}) \qquad (1)$$

where

$$a_o = \frac{1}{T} \int_{-T/2}^{T/2} x(t)\, dt$$

$$a_n \atop n \geqslant 1 = \frac{2}{T} \int_{-T/2}^{T/2} x(t) \cos \frac{2\pi n t}{T} \qquad (2)$$

$$b_n \atop n \geqslant 1 = \frac{2}{T} \int_{-T/2}^{T/2} x(t) \sin \frac{2\pi n t}{T} \;\; .$$

The position of the time axis can be chosen so that the mean value
(and thus a_o) is zero. The remaining coefficients a_n and b_n will, in
general, be different and their values may be plotted against frequency
as shown in figure 2.

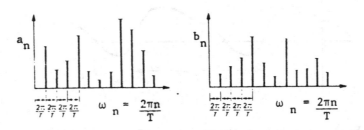

Figure 2

This gives a frequency domain description of the function in terms of
each of the amplitudes and frequencies of the sinusoidal components
(harmonics) that must be added together to synthesize the original
wave form. The location of the nth coefficient is at the frequency
of the nth harmonic, $\omega_n = \frac{2\pi n}{T}$.

Note that the spacing between adjacent harmonics is $\Delta\omega = \frac{2\pi}{T}$.
As T becomes large, $\Delta\omega$ becomes small and the coefficients a_n and b_n
become correspondingly packed more tightly. When the limiting condition
is reached, i.e. when $T \rightarrow \infty$, x(t) becomes aperiodic rather than a
periodic function in the time domain while in the frequency domain the
discrete Fourier coefficients turn into continuous functions of frequency called Fourier transforms.

1.2 Aperiodic Functions.   (FOURIER TRANSFORM ANALYSIS)

To develop a form of the Fourier transform to treat aperiodic functions, we can begin by substituting equation 2 into equation 1, and the using the relations $\frac{2\pi n}{T} = \omega_n$ and $\frac{1}{T} = \frac{\Delta\omega}{2\pi}$ to obtain

$$x(t) = \sum_{n=1}^{\infty} \frac{\Delta\omega}{\pi} \int_{-T/2}^{T/2} x(t) \cos \omega_n t dt \quad \cos \omega_n t$$

$$+ \sum_{n=1}^{\infty} \frac{\Delta\omega}{\pi} \int_{-T/2}^{T/2} x(t) \sin \omega_n t dt \quad \sin \omega_n t \; . \tag{3}$$

In the limit as $T \to \infty$, $\Delta\omega \to d\omega$ and Σ becomes an integral with the limits $\omega = 0, \infty$ such that

$$x(t) = \int_{\omega=0}^{\infty} \frac{d\omega}{\pi} \int_{-\infty}^{\infty} x(t) \cos \omega t dt \quad \cos \omega t$$

$$+ \int_{\omega=0}^{\infty} \frac{d\omega}{\pi} \int_{-\infty}^{\infty} x(t) \sin \omega t dt \quad \sin \omega t \; . \tag{4}$$

Calling $A(\omega)$ and $B(\omega)$ the components of the Fourier transform of $x(t)$ where

$$A(\omega) = \frac{1}{2\pi} \int_{-\infty}^{\infty} x(t) \cos \omega t dt$$

$$\text{and } B(\omega) = \frac{1}{2\pi} \int_{-\infty}^{\infty} x(t) \sin \omega t dt \tag{5}$$

$$\text{gives } x(t) = 2 \int_0^{\infty} A(\omega) \cos \omega t \, d\omega + 2 \int_0^{\infty} B(\omega) \sin \omega t \, d\omega \tag{6}$$

Equation 6 is a representation of x(t) by a Fourier integral or inverse Fourier transform.

A complex form of the Fourier transform can be obtained from the above development by first making use of the result that exp iωt = cos ωt + i sin ωt and defining $X(\omega) = A(\omega) - i B(\omega)$. Combining these two relations with Equation 5 gives (after some manipulation)

$$X(\omega) = \frac{1}{2\pi} \int_{-\infty}^{\infty} x(t) \exp(-i\omega t) \, dt \tag{7}$$

In a similar manner, Equation 6 may be manipulated into the form

$$x(t) = \int_{-\infty}^{\infty} X(\omega) \exp(i\omega t) \, d\omega \tag{8}$$

Equations (7) and (8) are called Fourier transform pairs, i.e. $X(\omega)$ is (the complex) Fourier transform of x(t)

Finally, an example of the limiting process of arriving at the Fourier integral from the series representation is shown in Figure 3. Note that Figure 3d shows the final transient function in terms of the complex Fourier transform pair given in Equation (7) and (8).

1.3 Aperiodic Functions.   (SHOCK SPECTRUM ANALYSIS)

Shock phenomena occur frequently in the field of sound and vibration. They originate mainly when large amounts of energy are suddenly released over short periods of time as, for example, in metal punching machines, explosions, or the rough handling of equipment. A simple shock may be defined as a transmission of kinetic energy to a system which takes place in a relatively short time compared with the

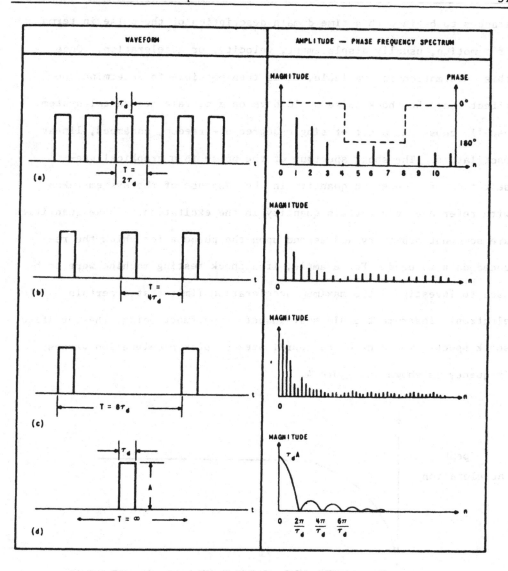

Figure 3 Transition of the Fourier line spectrum of a train of
         pulses to the continuous spectrum of an isolated pulse.

natural period of the oscillation of the system.

In the analysis of a shock pulse, it is essential in most instances to begin with a time domain description of the pulse in terms

of a motion, usually displacement, velocity, or acceleration. Once

this information is available, it is then possible to determine the

effect that the shock pulse would have on a certain mechanical system,

usually chosen as a set of single-degree-of-freedom, undamped, linear

oscillators. The shock spectrum of the pulse is a graphical pre-

sentation of a selected quantity in the response of the system taken

with reference to a certain quantity in the excitation. These quantities

are somewhat arbitrary and depend upon the purpose for which the re-

duced data is used. For example, if a shock testing machine were to be

used to investigate the maximum acceleration limit that a certain

electronic instrument could sustain before malfunctioning, the specified

shock spectra would have the coordinates of peak acceleration verses

frequency as shown in Figure 4.

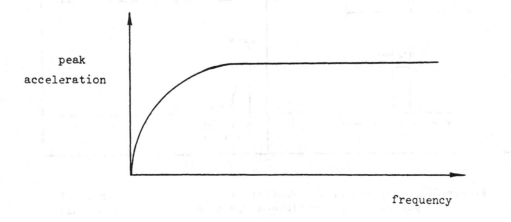

Figure 4

The testing machine would then have to produce a shock spectrum to

meet these specifications within an acceptable limit. Suppose for example, that the testing machine produced an actual acceleration at its base, $\ddot{Z}$ as shown in Figure 5.

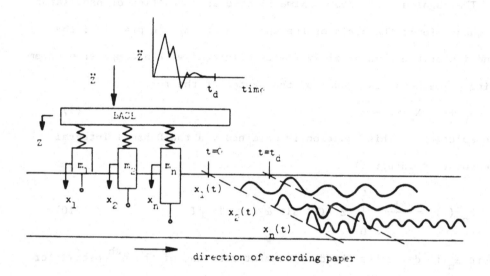

Figure 5

To illustrate how the shock spectrum of $\ddot{Z}$ (t) would be obtained, consider the above mechanical system consisting of a series of n un-damped, simple oscillators each having a resonant frequency one Hertz greater than its (n-1) neighbour. The oscillators are connected to the base of the testing machine via their springs. To each of the os-cillators is attached a pen to the record the displacement response of each oscillator during and after the impressed shock. The time history of the response is recorded by moving a strip of paper at an appropriate speed in the direction transverse to that of the pen. Note that in the above example there are two distinct parts of the response data, namely, the response during the shock $(0 \leq t \leq t_d)$ and the response

after the shock $(t \geq t_d)$. Within these two distinct periods the dis-

placement undergoes a maximum value (positive and negative).

The magnitude of these maxima plotted as a function of oscillator

frequency forms the basis of the shock spectrum. In practice, the

function of the mechanical system is simulated with a computer program

using a mathematical model of the system of the form

$$\ddot{X}_n + \omega_n X_n = - \ddot{z} \tag{9}$$

The solution to this equation is obtained via the Duhamel integral

(convolution integral) as

$$X_n(t) = \frac{1}{\omega_n} \int_0^t \ddot{z}(T) \sin \omega_n (t - T) \, dT \tag{10}$$

where $X_n(t)$ describes the displacement response of the nth oscillator

to the impressed acceleration shock, $\ddot{z}$. The velocity and acceleration

response of the nth oscillator can be obtained respectively by once and

twice differentiating equation (10) with respect to time. By evaluating

the Duhamel integral at each frequency ω_n, the response can be obtained

over the frequency range of interest. The response at each frequency

can then be examined for maxima (positive and negative) values. A

plot of these values against frequency is referred to as the initial

shock spectrum. Similarly, a plot of the corresponding maxima that

occur in the response after the shock is removed is referred to as the

residual shock spectrum. An example of the shock spectrum for an

idealized terminal-peak sawtooth function of peak acceleration A and

duration t_d is shown in figure 6.

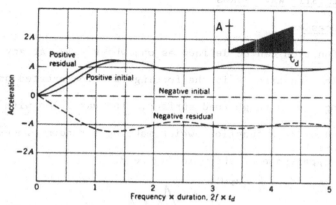

Figure 6

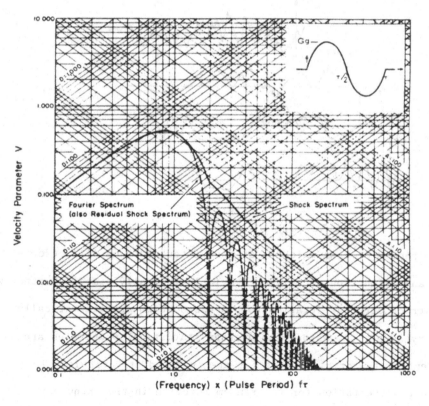

Figure 7   Fourier and shock spectra for full-cycle sine
             acceleration pulse

2. NON-DETERMINISTIC WAVE FORMS

2.1 Random Processes

A random function can be defined as one whose value at any instant
is unpredictable. An example is the forcing function exerted on a
wheel travelling over a rough road surface. The variation with time of
a set (ensemble) of such functions which make up a random process
might have the appearance as shown in figure 8.

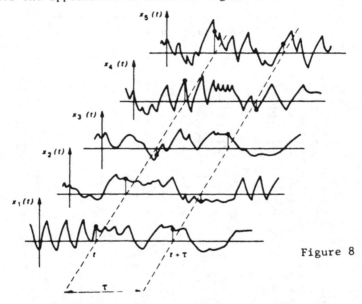

Figure 8

The random process which is of most interest to us will be continuous and a function of only one variable (normally time). Further we
will restrict our consideration to processes which are statistically
stationary, i.e., the associated average statistical properties are
independent of the origin chosen for the independent variable. To
develop a representation for the random function in the frequency
domain, we will need to extend the method of the Fourier transform to

include the class of non-deterministic (random) functions being considered. Since a random process x(t) has no beginning or no end, it is
not possible to evaluate the Fourier transform of x(t), since the integral involved would be unbounded. However, this difficulty can be
overcome by transforming an indirect time description of the function
which is bounded, namely, the auto correlation function. Let us
first examine some of the properties of this function.

2.2 Auto correlation Function

We define the auto correlation function for a random process x(t)
as the average value of the produce x(t) x(t+τ). The stationary process is sampled at time t and then again at time t + τ as shown in
figure 8. The value of the product $E[x(t) x(t + \tau)]$ will depend only
on the time separation so that we can write

$$E[x(t) x(t + \tau)] = R_x(\tau) \tag{11}$$

where $R_x(\tau)$ denotes the auto correlation function of x(t). A few of
the more important properties of this function are listed as follows:

(a) If x(t) is stationary, the mean of x(t), $E[x(t)]$ and the standard
deviation of x(t), $\sigma_x(t)$ will be independent of t so that

$$E[x(t)] = E[x(t + \tau)] = m \tag{12}$$

and $$\sigma_{x(t)} = \sigma_{x(t + \tau)} = \sigma \tag{13}$$

(b) The value of the auto correlation function can never be greater
   than the mean square value of x(t), $E[x^2(t)] = \sigma^2 + m^2$ and it can
   never be less than $\sigma^2 - m^2$.

(c) When the time interval separating the two points is zero, then

$$R_x(\tau = 0) = E[x^2(t)] \tag{14}$$

(d) For a stationary process, $R_x (\tau)$ depends only on the separation time

τ and not on absolute time t.

$$R_x(\tau) = E[\; x(t) \; x(t + \tau)] \;\; = \;\; E \; [\; x(t) \; x(t - \tau)] = R_x(-\tau) \quad (15)$$

so that $R_x (\tau)$ is on even function of τ.

The above properties are shown on the following graph of R_x versus

separation time τ.

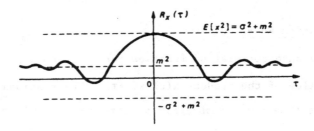

Figure 9   Illustrating properties of the autocorrelation
            function $R_x(\tau)$ of a stationary random process x(t)

As shown above, the auto correlation function describes certain

properties of the non-deterministic random process as a function of the

separation time τ. This function is deterministic and gives information

about the frequency content of the random process indirectly. This is

illustrated in figure 9 where the function $R_x(\tau)$ reflects the fre-

quency content of the sample functions of a random process x(t).

Thus, if the auto correlation function in the separation time domain is

Fourier transformed, a representation of the random process in the

frequency domain can be obtained. This representation is called the

mean square spectral density, $S_x (\omega)$ and is related to $R_x (\tau)$ through

the Fourier integrals

$$S_x(\omega) = \frac{1}{2\pi} \int_{-\infty}^{\infty} R_x(\tau) \exp(-i\omega\tau) \, d\tau \tag{16}$$

$$\text{and } R_x(\tau) = \int_{-\infty}^{\infty} S_x(\omega) \exp(i\omega\tau) \, d\omega \, . \tag{17}$$

A graphical form of the above Fourier transform pair is shown for several random functions in figure 10.

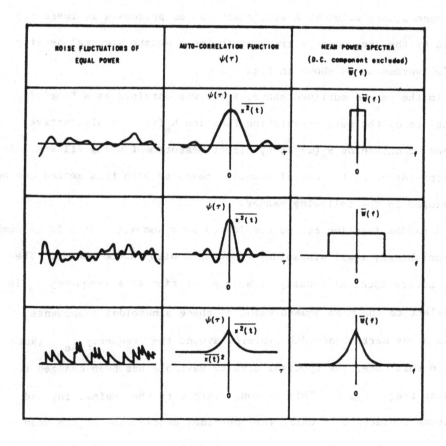

Note that if $\tau = 0$, equation 17 becomes

$$R_x(0) = \int_{-\infty}^{\infty} S_x(\omega) \, d\omega \qquad\qquad (18)$$

which, from the fundamental definition of $R_x(\tau)$, gives

$$E[\, x^2 \,] = \int_{-\infty}^{\infty} S_x(\omega) \, d\omega \qquad\qquad (19)$$

The mean square value of a stationary random process x is therefore given by the area under a graph of the mean square spectral density $S_x(\omega)$ versus ω, as shown in figure 10.

In the method outlined above, $S_x(\omega)$ was obtained as a Fourier transform of the auto correlation function $R_x(\tau)$. An alternative method of obtaining $S_x(\omega)$ is by direct measurement using filters. An interpretation of the actual quantity measured with this method can be developed in the following manner.

A random function can be considered as a summation of a large number of vanishingly small sinusoidal components with random phases. The mean square spectral density of such a function at a frequency f_o is equivalent to the mean square value of these sinusoidal components within a one Hertz bandwidth centered around the frequency f_o. (Note that in this case, the spectral density variable has been changed to circular frequency, f. This is consistant with the engineering and measurement practice of using the one-sided description of the mean square spectral density function, denoted with the symbol W(f)).

For a continuous frequency distribution from $f = 0$ to $f = \infty$ the mean square value of the random function is equal to the integral

$$E\ [x^2]\ =\ \int_0^\infty W(f_o)\ df_o \qquad\qquad (20)$$

which is identical to that given in equation (19).

Thus as shown in figure 11, a plot of the mean square spectral density $W(f)$ of a random function is a continuous curve and, at a frequency f_o, the power spectral density is represented by the average height of a rectangle of unit width centered about the frequency f_o. Figure 11 also shows a comparison of the spectrum of a wide and narrow band random signal. As stated earlier, the mean square value of these random signals is simply the area under the frequency spectrum plot $W(f)$ times the graphical scale factor.

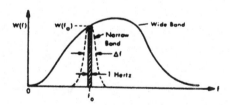

Figure 11   Wide Band and Narrow Band Spectra

REFERENCES

1. NEWLAND, D.E. 1975 Longman An Introduction to Random vibration and spectral analysis.

2. BENDAT, J.S. and PIERSOL, A.G. 1966 John Wiley, New York Measurement and Analysis of Random Data

3. BROCH, J.T. Bruel and Kjaer Mechanical Vibration and Shock Measurements.

4. The Shock and Vibration Bulletin April 1966 US Naval Research Laboratory Washington, DC. Bulletin 35 Part 6.

5. MORROW, C.T. 1963 John Wiley and Sons Inc Shock and Vibration Engineering.

THEORY OF ACOUSTICS I
Waves, Rays, and Statistical Fields

The following two lectures on the theory of acoustics are intended
to provide the scientist whose discipline is not primarily acoustics
and vibration with a brief review of the terminology and methodology
used throughout this advanced course. Emphasis will be mainly centered
on reviewing the methods available for defining wave motion in different
kinds of media under the influence of source perturbations at their
boundaries.

1. INTRODUCTION TO WAVE MOTION

The phenomenon of wave motion occurs in different media in many
different forms, e.g. waves on the ocean surface, vibrations on a
stretched string, electromagnetic radiation, sound propagation in the

air, and so on. When we ask what these various waves have in common,
at first we may be tempted to say that they all exhibit oscillation
or periodicity. Although this is frequently the case, it is not univer-
sal; the so-called tidal wave, for example, being just one exception. A
more characteristic property of wave motion emerges when observed from
a dynamic rather than a kinematic point of view. The arrival of a wave
at a certain point in the medium disturbs the matter in the vicinity of
that point and thus the wave must be regarded as transmitting energy from
one point to another. The transfer of energy by this means does not involve the transfer of matter, however, as no element of matter suffers
a permanent displacement from its original location. Rather, it moves
within some bounded region. The identification and quantification of
wave motion within this bounded region is the main occupation (and
headache!) of the acoustician. Theorists construct elaborate mathematical models to predict this motion for highly idealized conditions
while experimentalists and practitioners construct elaborate devices
to physically measure it without interfering too much with it.

Observed kinematically, different kinds of wave motion can be identified in terms of the particular paths an element of perturbed
matter describes when displaced within its bounded region. Expressing
this displacement relative to the direction of the wave front provides
a means of classification. The following table gives a few of the more
common wave motions encountered in acoustics along with their propagation velocities:

TABLE 1

WAVE MOTION	MOTION RELATIVE TO WAVEFRONT	MEDIUM	WAVE FRONT PROPAGATION VELOCITY
1. Compressional (longitudinal)	←•→	GASES	$[\dfrac{\gamma P_o}{\rho_o}]^{1/2}$
		LIQUIDS	$[K/\rho]^{1/2}$
		SOLIDS	$[\dfrac{E(1-\mu)}{\rho_o(1+\mu)(1-2\mu)}]^{1/2}$
2. Shear (transverse)	↕	SOLIDS	$[\dfrac{E}{2\rho_o(1+\mu)}]^{1/2}$
		STRINGS (c/s area A)	$[\dfrac{Te}{A\rho_o}]^{1/2}$
3. Flexural	∞	RECT. RODS (Thickness h)	$[\dfrac{Eh^2}{12\rho_o}]^{1/4}\omega^{1/2}$
		PLATES (Thickness h)	$[\dfrac{Eh^2}{12(1-\mu^2)\rho_o}]^{1/4}\omega^{1/2}$
4. Rayleigh	↻	SURFACES OF SOLIDS	$0.385\left[\dfrac{E(2.6+\mu)}{\rho_o(1+\mu)}\right]^{1/2}$

where
P_o = equilibrium pressure
γ = ratio of specific heats
K = isentropic bulk modulus of elasticity
ρ = mass density
E = extensional modulus (Young's modulus)
μ = Poisson's ratio
Te = tension
ω = angular frequency

Quantifying these different wave motions dynamically brings in the consideration of time and thus knowledge of the velocity and acceleration of the perturbed matter is required in addition to its displacement. In a gas, the medium which we will mainly be considering in this lecture, it is also necessary to quantify the changes in the state of the gas that result from the passage of a wave front. Thus a distinction must be made between the pressure, temperature, and density at a rest state and at a perturbed state. Throughout this lecture the following notation

will be used for the two sets of variables.

REST VARIABLES   PERTURBATION VARIABLES

pressure p_o displacement ξ
density ρ_o velocity v
temperature T_o acceleration a
                                        pressure p
                                        density ρ
                                        temperature T

The pressure, density, and temperature of the gas locally at any

instant are thus given by $p_o + p$, $\rho_o + p$, and $T_o + T$ respectively.

To describe a particular wave field, it is sufficient to specify

the position and time dependence of two of the perturbation variables.

For gases these are usually chosen as the pressure, p, and the velocity,

v. The ratio and product of these two variables define two important

dynamical properties of the medium and the wave motion perturbing it,

namely the impedance and intensity respectively.

Symbolically, the impedance (referred to as the specific acoustic

impedance) is given as

$$Z = \frac{p}{v} \tag{1}$$

while the intensity, which is the time average of the acoustic power

transferred per unit area, is

$$I = \overline{p\,v} \tag{2}$$

With the above definitions in mind, let us next develop briefly three

methods that are available for approaching problems in acoustics, namely,

wave acoustics, ray acoustics and energy acoustics.

1.1 WAVE ACOUSTICS

To develop this approach, we will begin by assuming that the gas

supporting the acoustic perturbation has the following properties; (1)
continuity, (2) homogeneity, (3) no internal friction, and (4) no thermal
conductivity. Further, the perturbations themselves will consist of
small amplitude displacements. If these idealizations are met to within
practical limits, the corresponding wave motion is governed by the wave
equation of the form

$$\left[\frac{\partial^2}{\partial x_i^2} - \frac{1}{c_o^2} \frac{\partial^2}{\partial t^2}\right] \begin{bmatrix} p \\ \xi_i \\ v_i \end{bmatrix} = 0 \qquad\qquad (3)$$

where $\partial^2/\partial t^2$ and $\partial^2/\partial x_i^2$ are second order partial derivatives with respect to time, t, and space coordinates, x_i, respectively. Note that
the wave equation can be expressed in terms of any of the perturbation
variables, p, ξ_i, v_i, etc. since these are all uniquely related to one
another. Solutions to the wave equation include the general form $p(x_i \pm c_o t)$ where $p(x_i)$ describes a pressure wave profile at the time t = 0.
For the one-dimensional case ($x_i = x$), $p(x - c_o t)$ describes the pro-
gression of the pressure wave profile in the positive x direction (a
forward propagating wave) while $p(x + c_o t)$ describes that in the negative
x direction (a backward propagating wave).

Example: One dimensional transverse wave motion on an infinite string

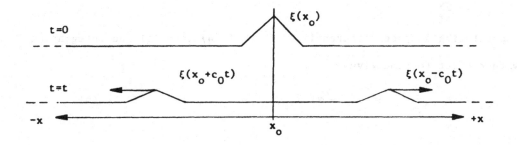

An especial kind of wave motion is that described by the harmonic, sinusoidal (or exponential) wave given as $\sin k(x \pm c_o t)$ for a fixed value of c_o. The wave number $k = \omega/c_o = 2\pi/\lambda$ is proportional to the angular frequency, ω, at which the wave repeats itself in time. It is inversely related to the wavelength, λ, the distance over which the wave repeats itself in space. Sinusoidal wave motion is of fundamental importance in acoustics since every function $p(x_i \pm c_o t)$ that occurs in practice can be represented by Fourier's theorem as a sum or an integral of sinusoidal functions.

1.1.1 PLANE WAVE PROPAGATION IN AIR

In one dimensional space ($x_i = x$) a forward moving sinusoidal pressure wave of amplitude $\hat{p}$ is given as

$$p = \hat{p} \sin \omega(t - x/c_o) . \tag{4}$$

With changing time the pressure distribution moves parallel to the x direction. At a particular time, t, the sound pressure is constant in planes perpendicular to the x direction, and thus the waves are referred to as plane waves. The impedance and intensity of plane waves can be determined by using Newton's second law (momentum equation) which relates p and v according to

$$\rho_o \frac{\partial v}{\partial t} = - \frac{\partial p}{\partial x} . \tag{5}$$

Substituting a space differentiated form of (4) into (5) and integrating with respect to time gives

$$v = \frac{\hat{p}}{\rho_o c_o} \sin \omega(t - x/c) . \tag{6}$$

The impedance of a plane wave is then given as

$$Z = \frac{P}{v} = \rho_o c_o \quad (kg \; m^{-2} \; sec^{-1}). \tag{7}$$

The product $\rho_o c_o$ is known as the characteristic acoustic impedance, the value of which in air at room temperature (20^oC) and standard pressure (1 atm) is 414 kg m^{-2} s^{-1}.

Being purely real (radiative) with no imaginary (reactive) component, this impedance represents physically the acoustic loading that would oppose the motion of a large plate vibrating as an interface between air and a vacuum.

The intensity of a plane sinusoidal wave is given as

$$I = \frac{1}{T} \int_0^T \frac{\hat{p}^2}{\rho_o c_o} \sin^2 \omega(t - x/c) dt \tag{8}$$

where T = 1 period. Hence.

$$I = \frac{1}{2} \frac{\hat{p}^2}{\rho_o c_o} = \frac{\overline{p^2}}{\rho_o c_o} \quad (W \; m^{-2}) \tag{9}$$

where $\overline{p^2}$ is the mean square pressure. In air, the mean square pressure corresponding to an intensity of 1 W m^{-2} is

$$\overline{p^2} = 414 \; (Pa)^2 \tag{10}$$

which gives a sound pressure level of

$$S.P.L. = 10 \; \log_{10} \frac{\overline{p^2}}{p_{ref}^2} = 20 \; \log_{10} \frac{\sqrt{\overline{p^2}}}{P_{ref}} \doteq 20 \; \log \frac{20}{20 \; \mu Pa} \doteq 120 \; dB \tag{11}$$

Two sinusoidal plane waves of equal amplitude travelling opposite directions combine to give a pressure field of the form

$$p = \hat{p}[\sin(\omega t-kx) + \sin(\omega t+kx)] = 2\hat{p} \sin \omega t \cos kx. \tag{12}$$

The resultant wave is called a standing wave wherein the pressure oscillates at a particular position with constant amplitude, but the

amplitude itself is a function of position. The corresponding velocity
is given as

$$v = - \frac{2\hat{p}}{\rho_o c_o} \cos \omega t \sin kx \qquad (13)$$

and thus the intensity of a standing wave is zero since the pressure
and velocity are 90° out of phase.

1.1.2 SPHERICAL WAVE PROPAGATION IN AIR

In considering three dimensional wave propagation, the most important type of wave is the spherical wave for which the sound field
quantities are functions only of time, t, and distance, r, from the
centre. An outgoing spherical harmonic pressure wave of amplitude $\hat{p}$
and angular frequency ω satisfies the spherical wave equation

$$\frac{\partial^2 (rp)}{\partial r^2} - \frac{1}{c_o^2} \frac{\partial^2 (rp)}{\partial t^2} = 0, \qquad (14)$$

with a solution of the form

$$P = \frac{P}{r} e^{i(\omega t - kr)}. \qquad (15)$$

The corresponding perturbation velocity is obtained from Newton's law
as

$$v = \frac{\hat{p}}{i\omega \rho_o} [\frac{ik}{r} + \frac{1}{r^2}] e^{i(\omega t - kr)} \qquad (16)$$

As in the plane wave, the velocity has only one component in the direction of propagation. Both pressure and velocity have constant amplitude
and phase on spherical surfaces centred on the origin.* This makes the
impedance of a spherical wave complex and dependent upon the ratio of
the radius, r, to the acoustic wavelength, λ, i.e.,

$$Z = \rho_o c_o [\frac{ikr}{1+ikr}] = \rho_o c_o [\frac{2\pi i (r/\lambda)}{1+ 2\pi i (r/\lambda)}] . \qquad (17)$$

Only when $r \gg \lambda$ does $Z \rightarrow \rho_o c_o$ in which case the associated pressure and velocity oscillate in phase and decrease inversely with r. When considering propagation from spherical sources, this rate of decrease is commonly referred to as the loss due to spherical spreading and in decibels represents a decrease in intensity of 6 dB per doubling of distance.

To compute the intensity of a spherical wave, the velocity can be written in terms of the real part of equation (16) in which both the radiative and reactive parts contribute as

$$v = \frac{\hat{p}k}{\omega \rho_o r} \cos(\omega t - kr) + \frac{\hat{p}}{\omega \rho_o r^2} \sin(\omega t - kr) . \qquad (18)$$

Writing the pressure in a similar manner yields an expression for intensity as

$$I = \frac{1}{T} \int_0^T \frac{\hat{p}^2}{r^2 \rho_o c} \cos^2(\omega t - kr) dt + \frac{1}{T} \int_0^T \frac{\hat{p}^2}{r^3 \rho_o c} \sin(\omega t - kr) \cos(\omega t - kr) dt \qquad (19)$$

The second integral is identically zero when T is equal to one period and thus the intensity becomes

$$I = \left(\frac{\hat{p}}{r}\right)^2 \frac{1}{2\rho_o c_o} = \overline{(p^2)} \frac{1}{\rho_o c_o} . \qquad (20)$$

Note that the intensity falls off as the inverse square of the distance for a spherical wave. The rate of energy flow from a source, W, is given by

$$W = I \, 4\pi r^2$$
$$= \left(\frac{\hat{p}}{r}\right)^2 \frac{4\pi r^2}{2\rho_o c_o} = \frac{2\pi \hat{p}^2}{\rho_o c_o} . \qquad (21)$$

As expected from energy considerations, the rate of energy flow through any surface surrounding the source is the same.

1.1.3 PROPAGATION LOSSES

In the previous sections, the acoustic wave motion was assumed
to propagate in an idealized medium which was free of losses due to
viscosity, heat conduction and molecular absorption.   For a wide variety
of problems, these assumptions are realized practically and the solutions obtained from the corresponding wave equation adequately describe the particular wave field under consideration. When wave propagation over long distances is considered, however, the loss effects
in the media become significant and must be accounted for in the overall problem. Basically, the losses inherent in a gas can be classified under three main mechanisms which irreversibly convert acoustical
energy into heat energy as follows:

(1) Viscous losses − due to the viscous shearing forces of a gas

which oppose the particle motions of a sound wave. The attenua-

tion constant due to viscosity, α_v, at a frequency, f, is given

approximately as (1 atmos, $20^{\circ}C$)

$$\alpha_v = 8.5 \times 10^{-8} \; f^2 \; dB/km. \tag{22}$$

(2) Heat conduction losses − due to heat transfer between adjacent

condensation and rarefaction regions within the gas. The attenua-

tion constant due to heat transfer, α_T, at a frequency, f, is

given approximately as (1 atmos, $20^{\circ}C$)

$$\alpha_T = 3.6 \times 10^{-8} \; f^2 \; dB/km. \tag{23}$$

---

*Note, however, that the velocity field is made up of a radiative
(real) part, and a reactive (imaginary) part which is 90° out of
phase with the pressure field.

(3) Molecular absorption - energy losses due to a relaxation phenomenon of the molecules of a gas which are excited into resonance by the passage of a sound wave. In air, the principal effect involves an interaction of water vapour molecules with the resonance of oxygen and nitrogen molecules so that the molecular absorption is highly dependent on the humidity content of air. The corresponding attenuation constant, α_{MOL}, is thus a function of the frequency of the sound wave, f, and the molecular relaxation frequency of the air, f_m, and for oxygen relaxation at $20^{o}C$, is given by

$$\alpha_{MOL} = \frac{0.056 \; f^2/f_m}{[1 + (f/f_m)^2]} \;\; dB/km. \tag{24}$$

In combination the losses due to viscosity and heat conduction give the classical attenuation constant, α_{CL}, i.e.

$$\alpha_{CL} = \alpha_v + \alpha_T \tag{25}$$

The functional dependence of α_{CL} and α_{MOL} on frequency and humidity for air (1 atmos, $20^{o}C$) is shown in Figure 2.

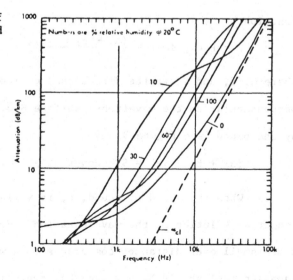

Fig. 2   Air Absorption
         Losses at $20^{o}C$.

1.2 RAY ACOUSTICS

In the previous section, the wave field was defined completely in functions involving time and space. This approach is feasible when used to describe wave propagation in unbounded (and certain bounded) media which are homogeneous. When considering wave propagation over large distances, as in the atmosphere, inhomogeneities due to wind, temperature gradients, etc. make the wave approach too cumbersome and a simplified approach must be adopted. Ray acoustics is one such simplification wherein families of rays are obtained as solutions of a simpler differential equation called the eikonal equation. In special cases, these solutions also satisfy the wave equation and under rather broad conditions provide good approximations to the more exact solutions.

In a stationary homogeneous medium (constant c_o and ρ_o), a solution to the three dimensional wave equation written in cartesian coordinates x, y and z can be expressed in the form of a sinusoidal function

$$P = \hat{P} \exp i\omega [t - \frac{(\alpha x + \beta y + \gamma z)}{c_o}] \qquad (26)$$

where α, β, and γ are the directional cosines of a straight line (ray*) perpendicular to the wavefront (surface S of constant phase) described by the plane (see figure 3)

$$(\alpha x + \beta y + \gamma z) = S = constant . \qquad (27)$$

When the speed of sound, c, is variable, equation (26) is no longer a solution of the wave equation. However, if the variations of c are small over distances comparable to a wavelength of sound, then equation (26) should be a reasonable approximation to a solution. One

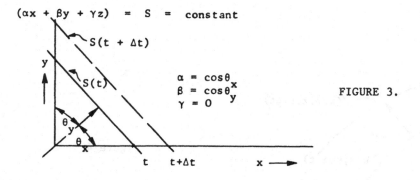

FIGURE 3.

immediate consequence of allowing c to vary with the space coordiantes is that α, β and γ are also variable and are no longer direction co-sines of the normal to the wave front. The wave fronts themselves are no longer planes but warped surfaces. Accordingly the rays are no longer straight lines but curved lines in space. To include these effects in a new sinusoidal function, it is mathematically convenient to express the frequency in terms of a constant reference velocity c_1 and its associated wavelength λ_1 so that the phase of the new function is given as

$$\exp i \frac{2\pi}{\lambda_1} [c_1 t - \frac{c_1}{c(x,y,z)}(\alpha x + \beta y + \gamma z)] . \tag{28}$$

The wavefront described by equation (28) is a surface on which the entire second term in the exponential has a constant value, and it propagates at speed c. Since c has different values at different locations along the wave front some portions of the wave move faster than others as shown in Fig. 4, where $c\Delta t$ is the perpendicular distance between two

_____

*However, it should be noted that in moving media, rays are no longer parallel to the local wave normal.

surfaces.

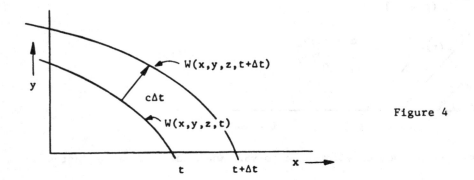

Figure 4

Note that a general function $W(x,y,z)$ has been adopted to describe the overall spatial contribution to the phase angle. Thus equation (28) can be rewritten as

$$\exp i \frac{2\pi}{\lambda_o} [c_1 t - W(x,y,z)] . \tag{29}$$

To write a trial solution to the wave equation which describes propagation in an inhomogeneous medium, the amplitude of the wave must be allowed to vary spatially in addition to the function, $W(x,y,z)$ such that

$$p = \hat{p}(x,y,z) \exp i \frac{2\pi}{\lambda_1} [c_1 t - W(x,y,z)] . \tag{30}$$

Taking the necessary derivatives of this solution and substituting them into the wave equation produces a complex expression which can be separated into real and imaginary parts. The real part is of most interest and is given as

$$(\frac{\partial W}{\partial x})^2 + (\frac{\partial W}{\partial y})^2 + (\frac{\partial W}{\partial z})^2 - \frac{c_1^2}{c^2} = \frac{\lambda_1^2}{4\pi^2 p} (\frac{\partial^2 \hat{p}}{\partial x^2} + \frac{\partial^2 \hat{p}}{\partial y^2} + \frac{\partial^2 \hat{p}}{\partial z^2}) . \tag{31}$$

Equating the left hand side of equation (31) to zero gives the eikonal equation. It can be seen that the function W which satisfies the eikonal equation is also a solution to the wave equation _if_ the

term on the right hand side of equation (31) is zero. This will be
true in general only in the limit of very high frequencies where $\lambda_1 \to 0$.
The approximation will be good, however, if this term is small compared
with $(c_1/c)^2$. Generally this condition is met whenever the change in the
sound speed is small over a local wavelength of sound.

The function $W(x,y,z)$ can be eliminated from the eikonal equation
to give three ordinary differential equations which together constitute
the equations of ray motion, namely,

$$\frac{d(n\alpha)}{ds} = \frac{dn}{dx} \; ; \quad \frac{d(n\beta)}{ds} = \frac{dn}{dy} \; ; \quad \frac{d(n\delta)}{ds} = \frac{dn}{dz}. \tag{32}$$

where ds is the arc length of the ray as shown in Figure 5.

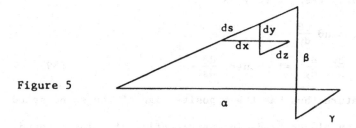

Figure 5

Note that $n(x,y,z) = c_1/c(x,y,z)$ is the index of refraction. The derivatives and directional cosines are illustrated in Figure 5.

To illustrate the usefulness of these equations, consider the
case where n varies only in the y direction such that

$$\frac{dn}{dx} = \frac{dn}{dz} = 0, \tag{33}$$

thus making

$$\frac{d(n \cos\Theta)}{ds} = 0 \tag{34}$$

and

$$\frac{d(n \sin\Theta)}{ds} = \frac{dn}{dy}. \tag{35}$$

From equation (34) it follows that $n \cos\theta$ has a constant value along a particular ray. If F and F' are two points on the ray, then

$$\frac{c_1}{c} \cos\theta = \frac{c_1}{c'} \cos\theta'. \tag{36}$$

Moreover, if F' is located at a reference point where $c'(y) = c_1$ and θ_1 is the direction of the ray at this point, equation (36) becomes

$$\frac{\cos\theta}{\cos\theta_1} = \frac{c}{c_1} = \frac{1}{n} \tag{37}$$

which is identical with Snell's law in optics.

The dependence of ray curvature on the change in sound speed can be obtained by writing equation (37) as

$$c = \text{const. } \cos\theta \tag{38}$$

and differentiating with respect to y as

$$\frac{dc}{dy} = -\text{ const. } \sin\theta \frac{d\theta}{dy}$$

$$= -\text{ const. } \frac{dy}{ds} \frac{d\theta}{dy} = -\text{ const. } \frac{d\theta}{ds}. \tag{39}$$

The ray curvature thus has the opposite sign of the sound speed gradient, i.e., the ray always bends in the direction of a lower sound speed region as illustrated in Figure 6.

Examples of ray curvature for various sound speed gradients are shown in Figure 7.

Sound speed gradients by wind are of special interest since the ray curvature is dependent upon the direction of the wind relative to that of the sound as shown in Figure 8.

Under the condition shown, the sound radiating from a source at position S to position A would be little affected by the wind velocity gradient while that radiating to position B would be greatly

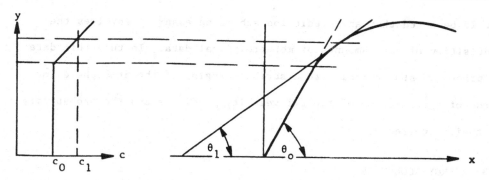

Figure 6

Examples of ray curvature for various sound speed gradients
are shown in Figure 7.

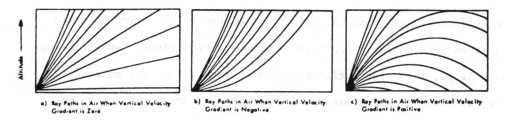

a) Ray Paths in Air When Vertical Velocity   b) Ray Paths in Air When Vertical Velocity   c) Ray Paths in Air When Vertical Velocity
   Gradient is Zero   Gradient is Negative   Gradient is Positive

Figure 7

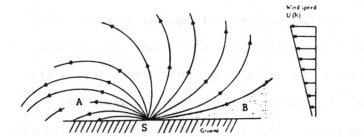

Figure 8

reduced due to the rays bending away from the ground surface. The region around position B is referred to as the shadow zone.

Although a discussion on the methods of measuring velocity gradients in the atmosphere is beyond the scope of these lectures, it

should be noted that any prediction scheme necessarily involves the
acquisition of vast amounts of meteorological data. In turn this data
is processed statistically to describe a region of the atmosphere in
terms of a collection of typical velocity gradients and the probability
of their occurrence.

1.3 ENERGY ACOUSTICS

The method of wave acoustics was discussed in section 1.1, where
the sound field was given in terms of exact solutions to the wave
equation. This same method can be applied to predicting sound fields
in relatively small enclosures of simple shape, by solving the wave
equation together with appropriate boundary conditions. When dealing
with large, irregular enclosures such as rooms, however, an exact
analysis is no longer possible and a method of descrbing the sound
field statistically must be adopted. Although such methods preclude
exact solutions, they do give fairly accurate average values of sound
quantities which are often all an acoustical engineer requires. Fundamental to this approach is the concept of the diffuse field in which
the sound energy emanating from a source in a room over a band of frequencies produces a sound field which is composed of a great many
superimposed rays propagating in all directions throughout the room with
equal probability. Moreover, the sound pressure level and thus the
energy density, E (joules/m^3), is constant throughout the volume of
the room. Under these conditions, the intensity, I, incident on any
imaginary surface in the room, is related to the local energy density
by

$$I = \frac{c_o E}{4} . \qquad\qquad (40)$$

The quantity I is the power incident per unit area of surface from all directions on one side of the surface. The energy approach (from which the Sabine equation derives) is developed by setting up an energy balance in the room between (1) the rate at which energy is absorbed at the boundary surfaces; (2) the rate at which energy contained in the air throughout the room changes; and (3) the rate at which energy is produced by the source. Consider each of these in turn.

First, each boundary surface in the room can be assigned an acoustical energy absorbing property called the sound absorption coefficient, α, which is a quotient of the sound energy that is not reflected and the incident sound energy. For an acoustically 'hard' wall, $\alpha=0$; for an open window, $\alpha=1$. In general, the sound absorption coefficient is a function of angle of incidence and frequency. Since the boundaries of an enclosure seldom consist of just one kind of material, it is necessary to represent α as an average of all the sound absorption coefficients, α_n, for all of the boundary surfaces, S_n, as

$$\alpha S = \sum_n \alpha_n S_n \quad (41) \qquad \text{where} \qquad S = \sum_n S_n . \qquad (42)$$

The quantity αS is usually written as a, which gives the total absorption present and has the dimensions of area. (Note that the averaging process described above is only valid where the absorbing surfaces are uniformly distributed along the enclosure boundaries). Recalling that the acoustic intensity has the units of energy per unit time per unit area, the product of a and the diffuse field intensity, I, gives the total energy absorbed per unit time by the combination of all the room's surfaces.

Next, an expression for the rate at which energy changes in the air throughout the room can be written in terms of a time derivative of the product of the room volume, V, and the energy density, E, as

$$\frac{d}{dt}[VE(t)] \tag{43}$$

which, in terms of intensity is (from equation (40))

$$\frac{4V}{c_o} \frac{d}{dt}[I(t)] . \tag{44}$$

Finally, using the above results, an expression for the energy in the enclosure containing a source of power, W, can be set up as

$$W(t) = a\ I(t) + \frac{4V}{c_o} \frac{dI(t)}{dt} . \tag{45}$$

If the time history of the source is such that

$$\begin{aligned} W(t) &= W_o \quad t < 0 \\ W(t) &= 0 \quad t > 0 \end{aligned} \tag{46}$$

the resulting solution to equation (45) gives an expression for the decay of the steady state sound field as

$$I(t) = \frac{W_o}{a} \exp(-\frac{ac_o t}{4V}) . \tag{47}$$

In this equation, the quantity, W_o/a, which has the units of watts per unit area, gives a measure of the steady state intensity (denoted by I_o) in the room. Note that this quantity is dependent upon room absorption as would be expected, but not on the volume or shape of the room.

For a diffuse field, the steady state mean square pressure is related to the energy density by

$$\overline{p^2} = \rho_o c_o^2 E \tag{48}$$

and thus the mean square pressure at any point in the room generated
by a steady state source of power, W, is

$$\overline{p^2} = \frac{4W\rho_o c_o}{a} .$$ (49)

Conversely, this expression can be used to determine the amount of absorption in a room provided that the power output of the source is known beforehand.

A more convenient method of determining absorption, however, is to use a transient decay measurement. This method makes use of equation (47) which, written in terms of intensity level, has the form

$$10 \log_{10} [\frac{I(t)}{I_o(t)}] = - \frac{4.34 \, ac_o t}{4V}$$ (50)

A measure of the decay rate is taken as the time required for the intensity, $I(t)$, to decrease by 60 dB from the initial steady state value, I_o. This time is called the reverberation time of the room and is given by

$$T = 0.163 \, \text{sec m}^{-1} \frac{V}{a} .$$ (51)

Consider an application of the above expression. Assume that a room measuring 10 m x 8 m x 5 m has surfaces with the following absorption coefficients:

	100 Hz		1000 Hz	
	α	αS (m^2)	α	αS (m^2)
Concrete floor	0.01	0.8	0.02	1.6
Brick walls	0.02	3.6	0.04	7.2
Tiled ceiling	0.50	40.0	0.80	64.0
		44.4		72.8
Reverberation time, T		1.5 sec.		0.9 sec.

In the above example, it can be seen that the reverberation time of the room is controlled mainly by the choice of absorptive material used for covering the ceiling.

REFERENCES

Section 1.1   KINSLER, L. and FREY, A. 1962 Fundamentals of Acoustics, Wiley and Sons, Inc.: New York, London.

MEYER, E., and NEUMANN, E.G. 1972 Physical and Applied Acoustics. Academic Press.
or Physikalische and Technische Akustik 1967 Friedr. Vreweg & Sohn G.m.b.H., Verlag, Braunschweig.

Section 1.1.3   EVANS, L.B., BASS, H.E., and SUTHERLAND, L.C. 1971 JASA 51, 1565-1575. Atmospheric absorption of sound: theoretical predictions.

Section 1.2   OFFICER, C.B. 1958 Introduction to the Theory of Sound Transmission. McGraw-Hill.

Section 1.3   KUTTRUFF, H. 1973 Room Acoustics. Applied Science Publishers Ltd., London.

THEORY OF ACOUSTICS II

Propagation and the Effects of Boundaries

The propagation of plane and spherical sound waves was discussed in Chapter 1 without reference to boundaries. In the present chapter we consider boundaries first as reflectors and then as sources of sound.

1. REFLECTION OF SOUND FROM PASSIVE BOUNDARIES

The sections which follow are concerned with reflection from rigid and nonrigid plane surfaces, and with the resulting interference between incident and reflected waves. The measurement of surface impedance in a standing-wave tube is discussed, and the first part of the chapter concludes with some practical considerations on the construction of sound-absorbent surfaces.

1.1   THE RIGID PLANE BOUNDARY: PLANE-WAVE REFLECTION AND INTERFERENCE

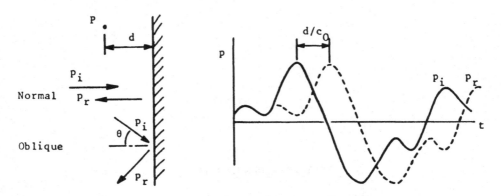

FIGURE 1.   Plane-wave reflection and interference (rigid boundary).

Figure 1 illustrates the process of interference between incident
and reflected plane waves at a rigid plane boundary of infinite extent.
Reflection occurs at the boundary with no change of amplitude or phase;
thus the velocity components normal to the boundary associated with the
incident and reflected waves cancel at the surface, as required by the
boundary condition.   The resulting interference at any point (e.g. P in
Figure 1) may be obtained by adding to p_i (the incident-wave pressure)
its value delayed by a time $2d/c_0$, which represents the round-trip time
from P to the boundary and back.

For oblique incidence at an angle θ (see Figure 1) the same applies,
except that the delay is $2d \cos\theta/c_o$.   This can be seen by considering the
plane-wave phase speed normal to the surface: for normal incidence the
phase speed is c_0, for oblique incidence it becomes $c_0/\cos\theta$.

The generalization to surfaces of finite impedance is greatly
simplified by considering waves of a single frequency, and is taken up in
section 1.4.   For the moment, however, we retain the rigid-surface

boundary condition in order to examine the interference which results from incident spherical waves.

1.2   THE RIGID PLANE BOUNDARY: SPHERICAL-WAVE REFLECTION AND INTERFERENCE

Figure 2 shows a source at Q radiating sound to an observer at P, with a rigid infinite "ground plane" at distance h below the source.

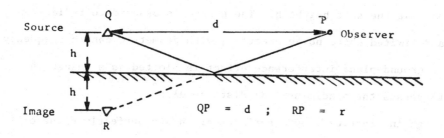

FIGURE 2.   Spherical-wave reflection and interference
(rigid boundary).

The boundary condition imposed by the ground plane, namely zero normal velocity over the entire surface, may be reproduced in the absence of the ground plane by adding an identical image source at R, i.e. in the mirror image position with respect to Q.   The effect of the reflecting surface is completely modelled, in the region above the plane, by the image source.

The pressure received at P may accordingly be written as

$$p = p_d + p_r, \qquad\qquad\qquad (1)$$

where p_d is the pressure which reached P directly from Q (i.e. as if the reflecting surface were absent), and p_r is the contribution from the image source at R.   To calculate p_r, it is sufficient to note that

(i)   The amplitude† is less by a factor d/r than that of p_d;

(ii)   the reflected signal suffers a time delay of $(r-d)/c_0$ relative
        to the direct signal;

(iii)   the radiation angle with respect to the source is different
        for the image and the actual source (see Figure 2); this is
        important where the source is directional. Note that the
        image source is inverted.

The nature of the resulting interference field is illustrared in

Figure 3, for the special case of an omnidirectional point source with

both Q and P at the same height h. The source is assumed to radiate an

octave band-limited white noise spectrum, with geometric centre frequency

f, and the ground-plane interference effect is plotted as a change in

level (ΔL) versus the nondimensional distance $d\lambda/h^2$.

Although in practice ground surfaces are never perfectly plane or

acoustically hard (especially at low frequencies), the above model does

demonstrate some of the problems inherent in measuring sound propagation

near the ground. When the ground impedance is finite, the reflection of

spherical waves cannot in general be accounted for by a simple image

source, and the analysis is correspondingly complicated. The reader is

referred to Delany and Bazley (1971) for a discussion of the resulting

interference phenomena, the most important being a substantial increase

in attenuation near the ground at large distances.

1.3   SCREENING OF SPHERICAL WAVES BY A BARRIER

The diffraction of spherical waves by a semi-infinite rigid plane is

another idealized problem whose solution is of practical interest, this

time in relation to barriers. As a method of noise control, barriers

─────────────

†Or r.m.s. value, in the case of time-stationary noise rather than
a single-frequency signal.

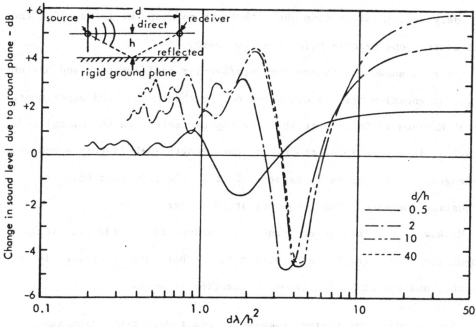

FIGURE 3.   Ground interference effect: hard surface, omni-
            directional point source.

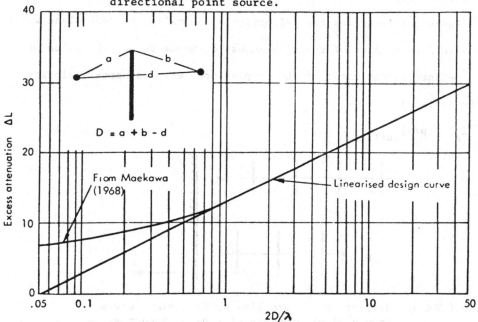

FIGURE4.   Excess attenuation due to a semi-infinite barrier.

are widely used in situations where there is no means of influencing the noise source, one example being roadway noise.

Figure 4 shows the source/observer/barrier configuration and presents barrier attenuation predictions based on the theoretical and experimental work of Maekawa (1968). Note that the key parameter, in this idealized situation, is the ratio of the excess path length (a+b-d) to the acoustic wavelength. When this ratio exceeds about 1, the predicted barrier attenuation increases with frequency at 10 dB per decade.

Maekawa (1968) has generalized his findings to provide practical methods for predicting the attenuation due to barriers of finite dimensions, and for including ground deflection effects.

1.4 PLANE BOUNDARY OF FINITE IMPEDANCE: PLANE-WAVE REFLECTION AND INTERFERENCE

Figure 5 illustrates the reflection of normally-incident plane waves from a uniform surface of specific acoustic impedance Z. The value of Z_s is defined as the ratio of the complex pressure and normal velocity at the surface:

$$Z_s = (p/v)_{surface}. \tag{2}$$

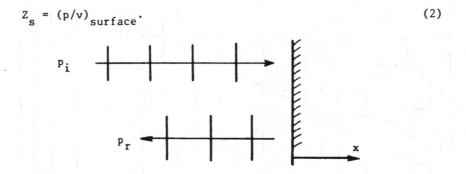

FIGURE 5. Reflection of normally-incident plane waves from a plane boundary of finite impedance.

Note that Z_s is in general a complex quantity, and refers to sound of single frequency; its value for any given boundary will be different at different frequencies. On the other hand, within the linear response range Z_s is independent of the surface pressure amplitude.

A further variable which can affect the specific impedance of a given boundary is the angle of the incident sound waves. The reason is that in principle, the normal velocity at one point on the surface is not determined just by the pressure at that point, but depends on the values of the pressure at all other points. A good example is the interface between two fluids. Provided - as here - we consider only one angle of incidence at a time, this does not matter; but the need to use a different Z_s value for each angle[†] should be borne in mind.

For the normal-incidence situation sketched in Figure 5, the pressure and velocity at any position x (measured normal to the boundary) may be written as

$$P = P_i + P_r; \qquad v = \frac{P_i}{\rho_o c_o} - \frac{P_r}{\rho_o c_o} , \qquad (3)$$

where v is taken as positive in the incident-wave direction. The ratio p/v at any value of x is therefore

$$\frac{P}{v} = Z(x) = \rho_o c_o (\frac{P_i + P_r}{P_i - P_r}) . \qquad (4)$$

If we denote the complex ratio of the reflected and incident pressures, p_r / p_i, by R(x) then equation (4) becomes

---

[†]A boundary for which Z_s is independent of the surface pressure distribution is sometimes referred to as a locally-reacting or point-impedance boundary.

$$\frac{Z}{\rho_0 c_0} = \frac{1 + R}{1 - R}. \tag{5}$$

The complex quantity R is the pressure reflection factor; its modulus r

is the reflection coefficient. The ratio $Z/\rho_0 c_0$, often denoted by ζ,

is the specific impedance ratio in the x direction. Equation (5) is

the basis of ~~the standing-wave tube method of impedance~~ measurement (see

section 1.5 below).

Problems

1. Show that for obliquely incident waves at angle of incidence θ,
    the impedance and reflection factor at any point in the standing-
    wave field are related by

    $$\zeta \cos\theta = \frac{1 + R}{1 - R} \quad , \qquad R = \frac{\zeta \cos\theta - 1}{\zeta \cos\theta + 1} \quad ; \tag{6}$$

    i.e. the generalization to oblique incidence simply involves
    replacing ζ by $\zeta \cos\theta$.

2. For normally incident waves as in Figure 5, show that the reflection
    factor varies with x according to

    $$R(x) = R(0) \exp 2ikx. \tag{7}$$

3. Show that the energy transmission coefficient in the direction normal
    to the surface, $\alpha_t = 1 - r^2$, is given under normal-incidence conditions
    by

    $$\alpha_t = \frac{4 \, \text{Re} \, \zeta}{|1 + \zeta|^2} \quad \text{(independent of x).} \tag{8}$$

The energy transmission coefficient defined above, which can refer

to plane waves at any angle of incidence, corresponds at a boundary to

the surface absorption coefficient defined in Chapter 1. Here it is

viewed as a property of an axial standing-wave field, not limited to

boundary surfaces.

1.5   THE IMPEDANCE TUBE

If a rigid-walled uniform tube is terminated with a sample of unknown acoustic impedance, the sample impedance value can be determined by exciting the tube with a single-frequency source and measuring the axial standing-wave pattern set up in the tube.

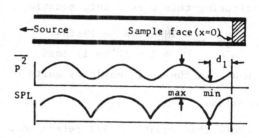

FIGURE 6.   Standing-wave pattern in an impedance tube.

Provided higher-order modes are excluded[†], the standing-wave field consists of incident and reflected plane waves travelling axially along the tube. Measurement of the corresponding reflection factor R_s at the sample face gives the normal-incidence impedance ratio through equation (5). For this purpose it is convenient to work with the amplitude and phase of R_s, as follows.

(i)   At the point labelled "min" in Figure 6, the pressures in the incident and reflected waves are 180° out of phase; at "max" they are in phase. The ratio of $|P_{max}|$ to $|P_{min}|$, denoted by N, is therefore related to the reflection coefficient $r(=|R_s|)$ by

$$N = \frac{1 + r}{1 - r} \; ; \qquad r = \frac{N - 1}{N + 1}. \tag{9}$$

(ii)   At "min" ($x = -d_1$), the phase of P_r relative to P_i is π radians; thus at $x = 0$, the relative phase is $\pi - \delta$ (where $\delta = 4\pi d_1/\lambda$).

[†]This implies an upper frequency limit of $f_{mas} = 0.586\, c_0/D$ (in a circular tube of internal diameter D), unless special precautions are taken.

It follows that the reflection factor at the sample face is

$$R_s = \frac{N - 1}{N + 1} \exp\, i(\pi - \delta). \tag{10}$$

Substituting this result into equation (5) gives finally

$$\frac{Z_s}{\rho_o c_o} = \frac{2N - i(N^2 - 1)\sin\delta}{N^2 + 1 + (N^2 - 1)\cos\delta}, \tag{11}$$

which allows the real imaginary parts of Z_s to be calculated from measured

quantities.

Note that equation (11) refers to a loss-free standing-wave tube, that

is one in which the incident and reflected waves suffer negligible atte-

nuation along the tube.   Corrections for tube losses are discussed by

Beranek (1949).

1.6 PHYSICAL REALIZATION OF ACOUSTICALLY ABSORBING BOUNDARIES

Acoustical design requirements

Typical applications of sound-absorbing surfaces are

(i) in a large enclosure, or room, to reduce the reverberant noise
     level;

(ii) in a duct, to absorb sound as it propagates along the duct.

In case (i), the sound field can often be treated as diffuse (see Chapter

1), and the most important acoustical characteristic of the surface is

its random incidence absorption coefficient $\bar{\alpha}$.

If the surface impedance Z_s is known for all angles of incidence

θ, the value of $\bar{\alpha}$ follows as

$$\bar{\alpha} = \int_0^{\pi/2} \alpha(\theta)\sin 2\theta \, d\theta, \tag{12}$$

where $\alpha(\theta)$ is the surface absorption coefficient at angle θ given by

$$\alpha(\theta) = \frac{4 \, \text{Re}(\zeta)\cos\theta}{|1 + \zeta\cos\theta|^2} \qquad \text{(compare equation (8)).} \qquad (13)$$

For example, if the surface is locally-reacting its ideal impedance

(for maximum $\overline{\alpha}$) is purely resistive, with $Z_s \doteq 1.5 \, \rho_o c_o$; $\overline{\alpha}$ is then 0.95.

In case (ii), however, the attenuation produced by a lined duct is

not uniquely determined by $\overline{\alpha}$, since we are not usually dealing with a

diffuse sound field - particularly if the dimensions of the duct cross-

section are not large compared with the wavelength. The optimum wall

impedance for maximum attenuation in a given length depends on the duct

dimensions, the frequency and the spatial characteristics of the sound

to be attenuated.

Porous materials for sound absorption

Figure 7 shows a type of con-

struction commonly used for sound-

absorbent surfaces. The porous

material may consist of rockwool

or fibreglass blanket, porous

material such as woodwool slabs

or acoustic plaster.

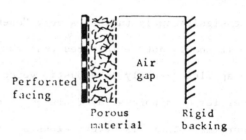

FIGURE 7. Components of a
sound-absorbing
boundary.

The function of the porous material is to allow the acoustic velocity

field to penetrate with relatively little obstruction until the energy

is dissipated by viscous action in the fine pores or interstices. As

a general rule, a thickness in excess of $\lambda/8$ (for material + air gap)

is required for optimum absorption. The aim is to achieve the largest

possible velocity amplitudes within the porous layer.

Additional energy dissipation is in principle provided by the thermal inertia of the fibrous or solid matrix, but this contribution is small compared with the viscous dissipation mentioned above, unless the total thickness is less than the $\lambda/8$ minimum.

In selecting porous materials for sound absorption, it may also be important to consider such features as flammability and mechanical integrity.

Significance of flow resistance for porous materials

For purposes of maximizing $\bar{\alpha}$, a useful guide is that the porous layer in Figure 7 should have a flow resistance† of around $3\rho_0 c_0$. If the layer is too thin, significant reflection will occur from the rigid backing; if the flow resistance is too high, the layer itself will reflect significantly (unless a very "open" material is being used).

It should not be inferred that the resistance measured for steady flow applies directly to acoustic velocity fluctuations. The acoustical properties of a porous layer are quite different from the steady-flow properties. There is nevertheless a close connection, for many porous materials, between the acoustical properties at a given frequency and the unit-depth flow resistance σ. Delany and Bazley (1969) found that their measurements of characteristic impedance and propagation constant, in a range of materials, collapsed well on the non-dimensional parameter $f\rho_0/\sigma$ for parameter values between 0.01 and 1.

---

†The flow resistance of a layer of porous material is defined as the ratio of the pressure drop across the layer to the volume flow rate per unit area through the layer, under conditions of slow steady flow.

Physically, the parameter $f\rho_0/\sigma$ -- or rather its square root -- is a measure of the ratio of the typical interstitial dimensions in a porous material to the viscous a-c boundary layer thickness.

The effect of varying the air gap

Increasing the air gap, in constructions of the type shown in Figure 7, generally improves the $\bar{\alpha}$ value up to a limiting depth of $\lambda/4$. An air gap of $\lambda/(2\cos\theta)$ (or integer multiple thereof) is equivalent to no gap at all. Thus an air gap chosen for low-frequency absorption can degrade the high-frequency absorption. In practice this is seldom a serious problem, since most porous layers absorb more effectively at high frequencies in any case.

Partitioning of the air gap into cells may offer advantages (including a closer approach to point-reacting behaviour); the partitions may have to extend into the porous layer as well to obtain the full effect.

The effect of facing layers

Little effect on the surface impedance results from perforated covers, provided the open area exceeds about 15% of the total. For situations where it is necessary to guard against fibre leakage or ingress of contaminants, finely-woven fabric facings are available with low flow resistance. The use of paint, or any other impervious coating applied directly to the surface of the porous layer, destroys the high-frequency absorption properties; lightweight membrane facings (e.g. plastic film) are permissible only if a gap is left between the facing and the porous layer, so that the facing can vibrate independently.

REFERENCES

BERANEK, L.L.   1949 Acoustic Measurements New York: Wiley.

DELANY, M.E. and ABZLEY, E.N.   1969 NPL Aero Report AC 37 Acoustical
    characteristics of fibrous absorbent materials.

DELANY, M.E. and BAZLEY, E.N. 1971 Journal of Sound and Vibration 16,
    315-322. A note on the effect of ground absorption in the
    measurement of aircraft noise.

MAEKAWA, Z.   1968 Applied Acoustics 1, 157-173. Noise reduction by
    screens.

The author wishes to acknowledge the assistance of Dr. C.L. Morfey
in the preparation of this chapter. Dr. Morfey is a senior lec-
turer at the Institute of Sound and Vibration Research, University
of Southampton, Southampton, England.

ACOUSTIC SOURCES I

Elementary Source Characteristics

In the previous chapters on the fundamentals of acoustics, the discussion on boundaries was limited to treating them as sound reflecting
or sound absorbing surfaces. In the chapters to follow, we will consider boundaries as sources of perturbations in the surrounding fluid
media. First we will look at the various mechanisms by which the perturbations are generated and describe the source characteristics in
terms of radiation impedance and acoustic power. Once the source characteristics are determined, we will describe the corresponding acoustic
field in the region containing the source in terms of the near and far
field, and the source directivity.

Perturbations to the equilibrium state of a fluid are brought abont by many different types of sources, the most important of which are:

    a)   Surface sources - the source consists of a vibrating motion of
        a surface which acts as a boundary to the fluid in which the
        sound propagates. Examples of this type of source are the mo-
        tion of a loudspeaker cone or a drum head.

    b)   Flow sources - in this type of source the perturbation acting
        as a source of acoustic waves is a secondary phenomenon accom-
        panying the flow of the media such as in jet flow, whistles,
        or the bubbling of brooks.

    c)   Explosion sources - the source results from a sudden expansion
        of the substance perturbing the surrounding media such as in
        the case of a bursting balloon or the growth and decay of ca-
        vitation bubbles in a fluid.

    d)   Thermal sources - perturbations to the surrounding media take
        the form of pressure and density changes that result from local
        changes in temperature such as in the combustion processes or
        thunder.

For purposes of analysing the source and field characteristics of
the above types of sources, it is convenient to note that all the sources
given may be reduced formally to equivalent surface sources. Such a reduction can be accomplished by first separating in the medium a closed
region containing the source and then determining the acoustic pressure
and perturbation velocity on the surface of that region which may be
treated as a surface source. Because all sources can be described in
this manner, we will confine the present discussion to surface sources
only.

Generally, an acoustic source may be considered from the following two points of view:

a) Source characteristics - properties of the source itself as a transmitter of acoustic energy in terms of

    i) acoustic power radiation

    ii) radiation impedance

b) Field characteristics - directional properties of the acoustic field produced by the source which depend upon

    i) the size and shape of source

    ii) The mode of vibration of the surface of the radiator, and

    iii) the presence of any large rigid surface near the vicinity of the source such as a baffle.

1. Source Characteristics

1.1 Acoustic power radiation

The power, W transmitted at a given instant across any closed surface S by acoustic waves is given by

$$W = \int_S pv_n ds \tag{1}$$

where p is the perturbation pressure caused by the acoustic wave at that instant and v_n is the corresponding component of the fluid perturbation velocity normal to the surface S. If the sound is being generated by the vibration of the surface of a solid in fluid (we assume that the fluid and the solid always remain in contact), we may take the surface S to be that of the vibrating solid and v_n will be the component of velocity of the solid normal to its own surface. Thus, in principle at least,

if the pressure on the surface of the vibrating solid and the corresponding normal component of velocity are known, the acoustic power radiation to which the vibration gives rise can be calculated.

In the absence of reflecting boundaries, the acoustic energy radiated away from a source represents energy of vibration which is removed from the source. The effect on the source is to reduce or dampen
its vibratory motion. This reaction on the source is known as acoustic
damping and its effect on the response of a structure to an applied
force is similar to that of the internal damping of the structural material itself. The acoustic damping may well be of the same order as
or greater than the structural damping. In contrast to this permanent
removal of energy from the source, reactive effects in the near field can
result in energy flowing into the pressure field during one interval
and being returned to the source in the next; this energy is of course
not radiated. Both the radiation (acoustic damping) and reactive effects
are taken into account in the radiation impedance of the source.

1.2 Radiation impedance

Earlier it was stated that if the pressure at and the normal velocity of a vibrating surface were known, the acoustic power output resulting from the vibration could be calculated. In very simple cases
which approximate to idealised models these quantities can often be calculated and the value of the radiated power obtained. In practice where
more complicated sources are encountered, such calculations are not possible. In such cases, however, the use of the idea of radiation impedance will often prove helpful in making rough estimates of radiation,
using the results of simpler cases as a guide.

The quantity radiation impedance, $Z_r = F/v_n$ (kg s^{-1}) is defined as the ratio of the force F (in newtons) exerted by the radiator on the medium to the normal velocity, v_n (m s^{-1}) of the radiator. The force is due to the reaction acting on the radiator given by $\int p \, dS$ when p is the acoustic pressure acting on the surface S of the radiator. The total impedance acting on a radiator is therefore the sum of its mechanical impedance Z_m and the radiation impedance Z_r defined above. Since these impedances are functions of frequency, ω, the velocity amplitude $v_n = \dfrac{F}{(Z_m + Z_r)}$ will not remain constant if the frequency is varied.

If we take p and v_n to have a simple harmonic variation and represent them as complex quantities, the time averaged complex power generated by the source must be represented as

$$\overline{W} = \int_S p \, v_n^* \, ds \tag{2}$$

where S is the surface of the solid and the asterisk indicates the complex conjugate of a quantity. The real part of $\overline{W}$ is the time-averaged radiated acoustic power, while the imaginary part is the reactive power. We can rewrite equation (2) as

$$\overline{W} = \int_S \overline{\frac{p}{v_n} v_n v_n^*} dS$$

$$= \int_S \overline{Z|v_n|^2} \, dS \tag{3}$$

where Z is the normal specific acoustic impedance of the vibrating surface, defined as the ratio of acoustic pressure at the surface to the component of velocity normal to the surface. We now define radiation impedance, Z_r as

$$Z_r = R_r + iX_r \qquad (4)$$

$$= \int_s \frac{Z \overline{|v_n|^2}}{v_c^2} \, dS \qquad (5)$$

where v_c is some characteristic velocity of the vibrating system. R_r and X_r are the radiation resistance and radiation reactance respectively. The radiated power is $R_r v_c^2$ and the reactive power is $X_r v_c^2$.

2. Field Characteristics

Sources which radiate sound uniformly in all directions are descri-bed as omni-directional sources, the simplest example being the pulsating sphere. Generally such sources are small in size compared to the wave-length of sound they radiate. Most practical sources, however, are made up of extended surfaces and the associated wave field is almost always directional, i.e. the radiated sound pressure levels are higher in some directions than in others. Directivity is a measure of the differences in radiated levels with direction at a given frequency and it is usually stated as a function of angular position around the acoustical center of the source. A plot of these levels in polar fashion at angles for which they were obtained is called a directivity pattern of the source.

In order to obtain a reference frame for a given source a principle axis is chosen which, as a rule, is along the direction in which the greater part of the energy of the source is radiated. It is along the principle axis that the quantity directivity factor is defined. At a given frequency, the directivity factor is the ratio of the square of the radiated free-field sound pressure at a fixed field point on the principal axis to the mean square sound pressure averaged over a sphere

passing through the fixed point and concentric with the acoustic center
of the source.

3. Examples of Spherical Sources

The various acoustical quantities discussed in the previous section
were introduced to describe the behaviour of surface sources and their
associated sound fields. Let us next examine two basic spherical sources
in detail to illustrate how these quantities may be used to provide useful approximations to the more complicated sources met in practice. To
develop this discussion, we will relate the acoustic pressure and velocity terms through the acoustic velocity potential function, ϕ in the
following manner.

$$p = \rho_o \frac{\partial \phi}{\partial t} ; \qquad v = -\frac{\partial \phi}{\partial r} \qquad\qquad (6)$$

where the coordinate r is taken in radial direction away from the center
of the source.

3.1 Pulsating Sphere (Monopole)

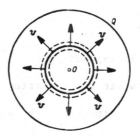

$$v_a = -(\partial \Phi / \partial r)_{r=a}.$$

Spherical sources belong to the simplest type of sources where the
wave field is described by surfaces of equal amplitude and phase which
have the shape of concentric spheres. If a 'pulsating sphere' is located at the center of a spherical wave field as a source, the field

distribution will not be changed. In this case the sphere is assumed to
vibrate such that its radius varies sinusoidally around a mean value a.
If the velocity of the radial vibrations of the sphere is $v_a = \hat{v} \exp i\omega t$,
the component of the acoustic velocity potential on its surface is
given by

$$v_a = -\left(\frac{\partial \phi}{\partial r}\right)_{r=a} .$$ (7)

If we write the acoustic velocity potential for a spherical wave propagating outward from the center at an amplitude A, and a frequency ω in
the form

$$\phi(r,t) = \frac{A}{r} \exp i(\omega t - kr + \theta)$$ (8)

the amplitude and phase respectively at $r = a$ are found to be

$$A = \frac{a^2 \hat{v}}{1+ika} \quad ; \qquad \qquad \epsilon = ka$$ (9)

and thus the acoustic velocity potential at the surface of the sphere
becomes

$$\phi = \frac{a\hat{v}}{1+ika} \exp i\omega t.$$ (10)

The acoustic pressure at the surface can then be written as

$$p = \rho_o \frac{\partial \phi}{\partial t}$$

$$= \frac{\rho_o a\hat{v}}{1+ika} i\omega \exp i\omega t .$$ (11)

We can now write the radiation impedance Z_r at the surface of the
pulsating sphere as the ratio the sound pressure to perturbation velocity in the form

$$\frac{Z_r}{4\pi a^2} = \frac{\rho_o c_o ika}{1+ika} = \rho_o c_o \left[\frac{k^2 a^2}{1+k^2 a^2} + \frac{ika}{1+k^2 a^2}\right].$$ (12)

When the diameter of the sphere is large compared to the corresponding wavelength of sound (ka >>1), the real part of the radiation im-

pedance, the radiation resistance R_r, is approximately equal to the radiation resistance of a piston radiator because each surface element of
a pulsating sphere acts somewhat like a piston radiator. For low frequencies or small spheres (ka <<1) the radiation impedance per unit surface is proportional to the square of the frequency or diameter of the
sphere, respectively. In summary,

$$\frac{R_r}{4\pi a^2} = \rho_o c_o \left[\frac{k^2 a^2}{1+k^2 a^2}\right] \tag{13}$$

$$\doteq \rho_o c_o \quad \text{if } ka >> 1$$

$$\doteq \frac{\rho_o}{c_o} \omega^2 a^2 \quad \text{if } ka << 1 .$$

The imaginary part of equation (12), the radiation reactance, is

$$\frac{X_r}{4\pi a^2} = \frac{\rho_o c_o ka}{1+k^2 a^2} . \tag{14}$$

The positive value of the reactance term means that the spherical surface is loaded by an inert mass. At high frequencies (ka >>1), the mass
per unit surface vanishes while at low frequencies (ka <<1)

$$\frac{X_r}{4\pi a^2} \doteq \rho c ka = \omega \rho a . \tag{15}$$

At low frequencies, therefore, the spherical surface acquires an additional mass of surface density ρa. The total mass of the medium that
vibrates along with the sphere at low frequenices is $4\pi a^2 \rho a$ which is
three times the mass $4/3 \pi a^3 \rho$ of the medium displaced by the radiator.
Thus we can see that the co-vibrating mass can be quite large at low
frequencies. The real and imaginary components of the radiation impedance of a pulsating sphere are shown in figure 2 as functions of ka.

The radiation resistance R_r multiplied by the square of the effec-

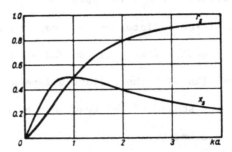

Figure 2   The components of the reduced specific impedance
r_s and x_s of a pulsating sphere

tive velocity, v_c^2 represents the power that is generated per unit area

by the sound source. For the case of the pulsating sphere, $\hat{v}$ is the

velocity amplitude at the surface and thus $v_c^2 = \hat{v}^2/2$. The corresponding

time averaged radiated sound power is

$$\overline{W} = R_r \; \hat{v}^2/2 = 2\pi a^2 \rho_o c_o \; \hat{v}^2 \; \frac{k^2 a^2}{1+k^2 a^2} \; . \qquad (16)$$

The strength of a pulsating sound source can be expressed in terms

of the volume flow it generates. This volume flow Q, can be defined

as the product of the surface area of the source and its velocity, $\hat{v}$ as

$$Q = 4\pi a^2 \; \hat{v} \; . \qquad (17)$$

Recall that the pressure at any field point r due to a pulsating source

of radius a can be written by

$$p = \frac{i\omega \rho_o a^2 \hat{v}}{(1+ika)r} \; \exp i \; (\omega t - kr + ka) \qquad (18)$$

or in terms of volume flow

$$p = \frac{i\omega \rho_o Q}{(1+ika)4\pi r} \; \exp i(\omega t - ka + ka) \qquad (19)$$

If the diameter of the sound source is smaller than one third of the

wavelength (ka <<1), eq. (19) can be simplified to

$$p = \frac{i\rho_o c_o kQ}{4\pi r} \; \exp i(\omega t - kr) \; . \qquad (20)$$

It can now be see that the sound pressure of a small spherical source is determined by its volume flow, and by nothing else. The geometry of a small source will not affect the pressure it generates at a sufficiently great distance from it. Thus sources producing equal volume flow will generate the same sound pressure and the same sound energy. This fact leads to considerable simplifications of the theory of sound radiation of small sources. Any two sources that have the same surface area and velocity amplitude generate the same volume flow. Any radiator of surface area S generates the same sound pressure and sound power as a sphere of the same area if both radiators have the same volume flow. Since they have the same area and generate the same sound power, they necessarily have the same radiation resistance per unit area and the same total radiation resistance. The radiation resistance per unit area of any small source is, therefore, the same as the radiation resistance per unit area of a sphere with the same area S, i.e.

$$\frac{R_r}{4\pi a^2} = \rho_o c_o \, k^2 a^2 = \rho_o c_o \, k^2 \, \frac{S}{4\pi} \, . \tag{21}$$

3.1.1 The monopole as a point source

Before going on to more complicated sources, it should be noted that the field produced by a pulsating sphere as its radius approaches zero represents an elementary solution to the wave equation called a point source. This solution has the form of equation (20) and describes the radiation generated at a mere point singularity. We will require such a soulution in our treatment of extended sources regions where

equivalent sources will be formulated in terms of a continuous distribution of such singularities.

3.2 Oscillating Sphere (Dipole)

A other basic type of spherical source is the so-called oscillating
sphere whereby the acoustic field is generated by the motion of a rigid
sphere oscillating forwards and backwards along its principle axis
(see figure 3.).

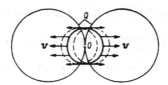

Figure 3 Radiation of a pulsating sphere

To determine the source and field characteristics of this source,
we will continue with the approach developed in the preceding sections
which involve the acoustic velocity potential. To begin, we assume that
a rigid sphere of radius a oscillates on its principle axis with a center
velocity v_o. The component normal to the surface which is decisive
for the acoustic radiation is

$$v_n = \hat{v}_o \cos \quad \exp(i\omega t) = - \left(\frac{\partial \phi}{\partial r}\right)_{r=a} \tag{22}$$

where θ is the angle between a line in the radial direction r and the
principle axis. The acoustic velocity potential for the oscillating
sphere can be shown to be given by

$$\phi = \frac{A}{r} \frac{1+jkr}{r} \cos \phi \exp i(\omega t - kr + \psi). \tag{23}$$

(Note that the cos θ dependence gives the corresponding directivity

pattern a shape of a figure eight.) Differentiating equation (23) and
substituting the result into equation (22) gives

$$A = \frac{a^3 \hat{v}_0}{2(1-ika)-(ka)^2} \quad ; \qquad \psi = ka \; . \tag{24}$$

Once the amplitude and phase are determined, the acoustic velocity potential can be used to obtain the normal velocity component and the
acoustic pressure on the surface of the sphere. Their ratio yields a
radiation impedance of the form

$$\frac{Z_r}{4\pi a^2} = (\frac{1}{3}) \frac{\rho_0 c_0 (ka)^4}{4 + (ka)^4} + i\rho_0 c_0 (\frac{ka}{3}) \frac{2 + (ka)^2}{4 + (ka)^2} \tag{25}$$

The co-vibrating mass of the sphere is

$$X_r = \frac{4}{3} \rho_0 \pi a^3 \frac{2 + (ka)^2}{4 + (ka)^2} \; . \tag{26}$$

Note that the radiation resistance as well as the co-vibrating mass of
the oscillating sphere are smaller than for a pulsating sphere of the
same diameter. For a small pulsating sphere it was shown that the radiation resistance is proportional to $k^2 a^2$; for a small oscillating
sphere, it is proportional to $k^4 a^4$. Thus whenever the sound source is
small or the frequency is low (ka <<1), the pulsating sphere component
of the sound energy is two orders of magnitude greater than the oscillating sphere component.

3.2.1 The dipole as an elementary solution

The next simplest elementary solution to the wave equation after
the monopole is the dipole, the field characteristics of which were
described above. In this case, the sound is generated by an injection
of momentum rather than mass into the surrounding fluid. An acoustic

dipole is equivalent to a force $\overline{F}$, concentrated at a point and varied in magnitude or direction or both. The strength of the dipole is equal to the force in terms of the time derivative of the acoustic velocity potential given in equation (23) in the form

$$p = \frac{i\omega\rho_o a^3 \hat{v} \ (1+ikr) \ \cos\theta}{r^2[2(1-ika)-(ka)^2]} \ \exp i(\omega t - kr + ka). \tag{27}$$

If we write the strength of the dipole as $\overline{F} = i\omega\rho_o 4/3 \ \pi a^3 \hat{v}_o$ and let a approach zero, we would obtain an expression for the field of a dipole given by

$$p = \frac{3 \ ik}{8r} \ \overline{F} \ (1+ \frac{1}{ikr}) \ \cos\theta \ \exp i(\omega t - kr). \tag{28}$$

ACOUSTIC SOURCES II

Extended Sources

In the preceding chapter we introduced two elementary solutions to
the wave equations, i.e. the monopole and dipole, and examined their
corresponding source and field characteristics. In the present chapter,
we will make use of these solutions in constructing equivalent source
representations of extended sources. Basically, the problem consists
of describing the acoustic field in a fluid medium produced by an extended source which consists of an arbitrary, vibrating surface S. To
illustrate the integral method of approaching this type of problem, we
will begin by assuming that the acoustic velocity potential ϕ, (α pressure) and its derivative normal to the surface $\partial\phi/\partial n$, (α velocity) are
specified on the surface boundary. Further we will assume that both
quantities have sinusoidal time variations at an angular frequency ω,

and that the surface S is a closed surface. (This last restriction,
however, can be relaxed simply by displacing part of the surface to
infinity). Once the boundary conditions at the surface are available,
they can then be combined with the appropriate elementary solutions to
give a solution for the acoustic field at point P in terms of an integral
given by

$$\phi\,(P) = \frac{1}{4\pi} \int_S [\frac{\exp(-ikr)}{r}\,\frac{\partial \phi}{\partial n} - \phi\,\frac{\partial}{\partial n}\,(\frac{\exp(-ikr)}{r})]\;dS \qquad (1)$$

where r is the distance between the point P and the element dS of the
surface. The above is the so-called Kirchhoff integral formula and has
the following physical interpretation:

The first term of the integrand represents the action of the vibrating surface treated as a continuous set of elementary point sources.
The source strength of each of these sources is proportional to the
component of the normal acoustic velocity ($v_n = -\partial\phi/\partial n$) and thus it
depends on the shape of the vibrating surface. The wave propagation
is in the accordance with Huygen's principle by which, each point of
the perturbation region may be considered as the origin of the perturbation having the form of a point source.

The second term in the integrand corresponds to the radiation due
to a distribution of dipoles along the surface. The strength of each
dipole is proportional to the acoustic pressure ($p = \rho_o i\omega\phi$), i.e. a
scalar quantity independent of the form of the radiating surface. The
wave field produced by each of the dipole sources varies with direction
as a cosine function.

The Kirchhoff integral can be simplified greatly if the vibrating boundary is plane and enclosed in an infinite baffle. In this case, the surface can be described in terms of either a layer of point sources only, or as a layer of dipoles, whichever is more convenient for the given problem. The corresponding integral is called Rayleigh's integral and has the form

$$\phi = \frac{\exp i\omega t}{2\pi} \int_S \hat{v}_n \frac{\exp(-ikr)}{r} dS .$$ (2)

Piston in a baffle

To illustrate the application of the integral method, consider the problem of a piston radiator in an infinitely large baffle. By this, we mean a plane surface that lies flush with an acoustically hard wall and vibrates in a direction normal to its surface in such a way that the velocity is constant in both magnitude and phase over the entire vibrating surface. If the piston is circular with a radius a (surface area S) and vibrates sinusoidally with a velocity $\hat{v}_n$ exp $i\omega t$, the corresponding wave field is given precisely by equation (2).

Source characteristics

If the above integral is evaluated for the problem being considered, the sound pressure and the perturbation velocity can be derived from the acoustic velocity potential in the usual manner. To calculate the radiation impedance of the piston it is necessary to find the force exerted by the piston on the fluid. The quotient of this reactive force and the piston velocity is the radiation impedance of the piston source in an infinite baffle. In this case, the radiation resistance and reactance have the form

$$R_r = \pi a^2 \rho_o c_o \; [1 - \frac{J_1(2ka)}{ka}\;] \tag{3}$$

$$X_r = \pi a^2 \rho_o c_o \; (\tfrac{4}{\pi}) \int_o^{\pi/2} \sin (2ka \cos \alpha) \sin^2 \alpha \, d\alpha \tag{4}$$

where $J_1(2ka)$ is a first order Bessel function. The behaviour of these
quantities as functions of ka is shown in figure 1. Also shown is the
radiation resistance and reactance for a circular piston in an infinite
tube. Note that above a ka value of two, the radiation resistance is
nearly the same for both cases.

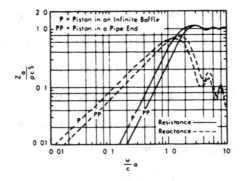

Figure 1   Acoustical resistance and Reactance per Unit
            Radiating Area, Normalized to the Characteristic
            Impedance of the Medium

Field Characteristics

It can be shown that the far field pressure generated by a circular piston in an infinite baffle is given by

$$p \doteq \frac{i \rho_o c_o ka^2 \hat{v}_n}{2r} \; \exp i(\omega t - kr) \; \frac{2J_1(ka \sin \theta)}{ka \sin \theta} \tag{5}$$

where r is the distance from the center of the piston to the field point.
The Bessel function divided by its argument gives the acoustic field a

'lobe like' directivity pattern as seen for various values of ka in
figure 2. Again the directivity patterns corresponding to a piston
in the end of an infinitely long tube are included for purposes of
comparison. Note that for both cases, the directivity pattern is
nearly omnidirectional for ka values of unity and below.

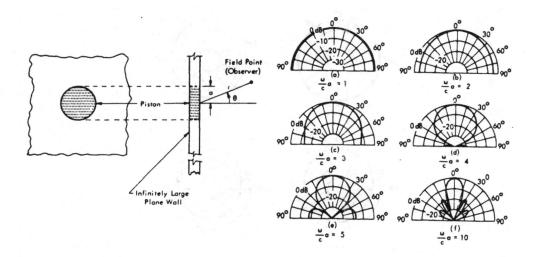

Figure 2a Directivity Patterns for a Rigid Circular Piston in
an Infinite Baffle

Application

The following is an application of the above development to a
practical problem.

Problem   A cover plate on a large machine is 0.3 m square and has a
fundamental natural frequency of 172 Hz. If it is vibrating pre-
dominately in this mode with a displacement amplitude at its centre of
0.25 mm, estimate the sound power radiated by the plate.

Solution: As an approximation we shall regard the vibrating plate as a

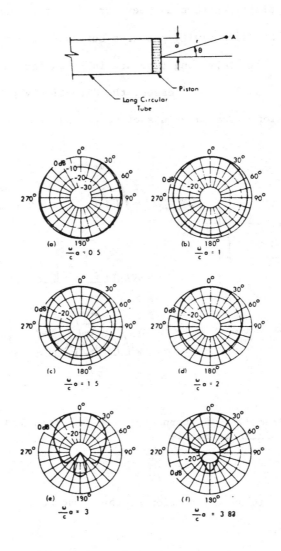

Figure 2(b)   Directivity Patterns for a Rigid Piston in
the End of a Long Tube

square piston in a large plane wall. It can be shown that, for a

square piston with a side length a enclosed in an infinite baffle, the radiation resistance is given by

$$R_r = a^2 \rho_o c_o [1 - \frac{2 J_1(ka)}{ka}]. \tag{6}$$

For this problem the constants can be evaluated as follows:

$$a = 0.30 \text{ m}$$

$$\rho_o c_o = 414 \text{ kgm}^{-2} \text{sec}^{-1}$$

$$\lambda = \frac{c_o}{f} = \frac{344}{172} = 2\text{m}$$

$$k = \frac{2\pi}{\lambda} = \frac{2\pi}{2} = 3.14 \text{ m}^{-1}$$

$$ka = (3.14)(0.30) = 0.94$$

$$J_1(ka) = J_1(0.94) = 0.42$$

Substituting the above results into equation gives

$$R_r = (0.30)^2(414) [1 - \frac{(2)(0.42)}{0.94}]$$

$$= 3.73 \text{ Nm}^{-1}\text{sec} \tag{7}$$

The velocity amplitude at the center of the plate is given by

$$\hat{v} = 2\pi f \times \text{displacement}$$

$$= (2\pi)(174)(.25)(10^{-3})$$

$$= 0.27 \text{ m sec}^{-1} \tag{8}$$

Using the mean square velocity at the center of the plate to calculate the average radiated power, we have

$$\overline{W} = R_r \frac{v^2}{2} = (0.373) \frac{(0.27)^2}{2}$$

$$= 0.136 \text{ watts} \tag{9}$$

The corresponding sound power level is

$$PWL = 10 \log_{10} \frac{0.136}{10^{-12}} \doteq 111 \text{ dB} \tag{10}$$

A better estimate of the power radiated would be obtained by assuming some shape for the vibrating fundamental mode and averaging the velocity over the plate.

Infinite Plate

In the preceeding section we saw how the Kirchhoff integral can be used to predict the sound fields produced by extended sources. In the example problem cited, the method was applied to predict the sound power radiated from a flat plate in an infinite baffle which was vibrating in a given mode shape. Another useful approach to approximating the source and field characteristics of extended plane surfaces that undergo periodic bending wave motion is the study of the infinite plate vibrating in a rectangular nodal line pattern of the form $v = \hat{v} \cos \kappa_x x \cos \kappa_y y$. In this model, it is assumed that the plate vibration is of a __fixed__ nodal line pattern, so that the corresponding structural wave numbers κ_s and κ_y in the x and y direction respectively are independent of frequency (see figure 3).

If the condition of continuity is satisfied at the surface of the plate, it can be shown that the acoustic wave number ($k = {}^{\omega}/c_o$) and the above structural wave numbers are related through the directional cosine of the sound waves n_z (see figure 3) that are generated by the plate by

$$k^2 = \kappa_x^2 + \kappa_y^2 + k^2 n_z^2$$

or

$$n_z = [1 - (\tfrac{\kappa}{k})^2]^{\tfrac{1}{2}} \tag{11}$$

where

$$\kappa^2 = \kappa_x^2 + \kappa_y^2 .$$

When $k = \kappa$, n_z is zero and the bending wavelength in the plate is equal to the acoustic wavelength in the surrounding medium. It follows that

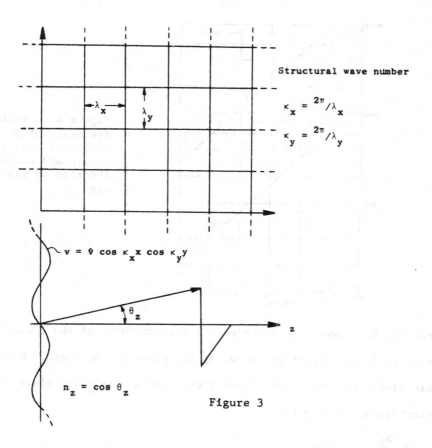

Structural wave number

$$\kappa_x = {}^{2\pi}/\lambda_x$$

$$\kappa_y = {}^{2\pi}/\lambda_y$$

$v = \hat{v} \cos \kappa_x x \cos \kappa_y y$

$n_z = \cos \theta_z$

Figure 3

the bending wave velocity for the plate c_B must equal the sound velocity in the medium c_o, and thus we define the frequency f_o, at which
this phenomenon occurs, i.e.

$$f_o = \frac{kc_o}{2\pi} = \frac{kc_B}{2\pi} \tag{12}$$

as the coincidence frequency of the plate. Examples of the coincidence
frequenices for a plate in air and water are shown in figure 4.

The sound pressure radiated from the vibrating plate has the form

$$p = A \cos \kappa_x x \cos \kappa_y y \exp(-ikn_z z) \tag{13}$$

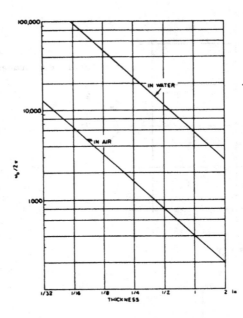

Figure 4. Coincidence
frequency of an alu-
minium or iron plate
in air and water, as a
function of its thick-
ness

where the term $\exp(-ikn_z z)$ represents the component of the acoustic

radiation in the direction normal to the plane of the plate. For har-

monic time dependence, the acoustic pressure gradient is related to the

perturbation velocity by

$$- \left. \frac{\partial p}{\partial z} \right|_{x=0} = i\omega\rho_o v \tag{14}$$

or

$$v = \frac{i}{k\rho_o c_o} \frac{\partial p}{\partial z} . \tag{15}$$

Combining a space differential form of equation (13) with equation (15)

yields an expression for pressure in the form

$$p = \frac{\rho_o c_o}{n_z} \hat{v} \cos \kappa_x x \cos \kappa_y y \exp(-in_z kz) \tag{16}$$

Source characteristics

Using the quantities developed above we can now determine the source characteristics of an infinite flat plate. The acoustic impedance per unit area of the plate is given by the ratio of the acoustic pressure and perturbation velocity at the surface of the plate by

$$z = \frac{p}{v} = \frac{\rho_o c_o}{n_z} = \frac{\rho_o c_o}{[1 - (\frac{\kappa}{k})^2]^{\frac{1}{2}}} = \frac{\rho_o c_o}{[1 - (f_o/f)^2]^{\frac{1}{2}}} \tag{17}$$

In the frequency range where $f/f_o < 1$, the acoustic impedance becomes a mass reactance:

$$z = \frac{i\rho_o c_o}{\left[(\frac{f_o}{f})^2 - 1\right]^{\frac{1}{2}}} = i \, 2\pi f m \tag{18}$$

where

$$m = \frac{\rho_o c_o}{2\pi \, (f_o^2 - f^2)^{\frac{1}{2}}} \tag{19}$$

is the effective mass per unit area of the plate.

Note that when $f = f_o$, i.e. at the coincidence frequency, the impedance becomes infinite. Above the coincidence frequency, the factor n_z is real and the acoustic impedance be purely a radiation resistance, i.e.

$$z = \frac{\rho_o c_o}{[1 - (f_o/f)^2]^{\frac{1}{2}}} \doteq \rho_o c_o \tag{20}$$

The impedance as a function of the frequency ratio, f/f_o is shown on figure 5. Note that the radiation resistance becomes practically equal to $\rho_o c_o$, the characteristic acoustic impedance of the medium, when the frequency is more than twice the coincidence frequency.

Field Characteristics

In the previous section on source characteristics, we examined the radiation impedance for various ratios of f/f_o which yielded real or imaginary values for the directional cosine n_z. We now want to look at

Figure 5 Radiation impedance
of an infinite plate excited
to a sinusoidal vibration
pattern; solid curve, radiation impedance; broken curve,
mass impedance

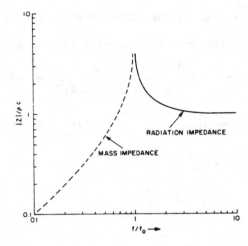

the corresponding acoustic field under these conditions. Recall that

the sound pressure field due to the vibrating infinite plate was given

as

$$p = \frac{\rho_o c_o}{n_z} \hat{v} \cos \kappa_x x \cos \kappa_y y \exp i(\omega t - n_z kz) \qquad (21)$$

At low frequencies, $f/f_o < 1$, n_z is then imaginary, and the cor-

responding sound pressure generated by the plate vibrations decreases

exponentially with distance from the plate:

$$p \doteq \frac{\rho_o c_o}{n_z} \hat{v} \exp \left(\frac{-2\pi z}{\lambda_o}\right) \cos \kappa_x x \cos \kappa_y y \exp i\omega t \qquad (22)$$

where $\lambda_o = c_o/f_o = \frac{2\pi}{\kappa} = \frac{2\pi}{k_o}$ is the characteristic wavelength.

Thus at low frequencies, the sound field is a 'wattless' near field with

no acoustic energy being radiated to great distances. Physically this

means that the distances between maxima and minima of the vibrator are

short enough so that the pressure variations over the surface can set

up a pure flow of fluid from the compressed regions to the rarefied regions. (Note that the exponential decrease of the near field with distance is governed by the characteristic wavelength of the plate and not by the sound wavelength).

At the condition $f = f_o$, $n_z = 0$ and the acoustic waves propagate in a direction parallel to the surface of the plate. All the sound energy generated by each region of the plate at infinite distances propagates parallel to the plate and contributes to the total sound field, thus making the sound pressure (and acoustic impedance) become infinite. In practice, of course, this condition is never met.

As the frequency increases above the coincidence frequency, $f > f_o$, the direction of the propagation turns away from the plate until, at high frequencies, the waves propagate normal to plate. In this frequency range the radiation impedance becomes equal to $\rho_o c_o$.

ACOUSTIC SOURCES III

Sources Due to Mechanical Impact

In our treatment of sources up to now, we assumed that surfaces which generate sound vibrate periodically either in a rigid body motion as in the case of the oscillating spherical source or, in a wave motion as in the vibrating plate example. When treating impacting surfaces as sources, the description of the motion of the two or more impacting surfaces is no longer as straightforward. The motion of the impactor prior to its collision with the impactee is mainly a rigid body one - simply a mass undergoing an acceleration (see fig. 1).

Examples of common impactors in industry would be hammers, punches, saw teeth, etc. These are mainly tools that are used for shaping or forming materials which assume the role of the impactee. Upon collision and immediately after, both the impactor and impactee can exhibit rigid body

prior to impact   after impact   Figure 1.

motion as well as wave motion. The former decays quickly while the latter decays more slowly depending mainly upon the amount of absorption
present in the impacted structure. The more slowly decaying wave motion
generates what is referred to as the 'ringing' sound following a mechanical impact.

   Needless to say, there is no exact or even approximate method
available for analysing complex impacting surfaces for the purposes of
predicting their sound radiating properties. From a noise control point
of view, the difficulties of analysing impact problems can be summarized
as follows:

1. On a complex machine such as a punch press, it is difficult to
    identify the surfaces of the impactor and impactee which radiate
    sound. In many cases, the surrounding machine components and indeed
    the floor supporting them become part of the impacted surface after
    collision. This feature complicates the radiation problem consider-
    ably.

2. The magnitude of the response (rigid body and wave motion) of the
    impacted surfaces depends upon (a) their mechanical (and acoustical)
    impedance and (b) the form of the energy present in the impactor.
    Even if the geometries of the impacting surfaces were well defined,

a measure of their acceleration response during and after impact
would be exceedingly difficult to obtain in practice.

3. As shown in the previous sections, the radiation resistance of each
   of the impacted surfaces is required for the calculation of the
   sound energy radiated by the entire mechanical system. Analytical
   expressions for this frequency dependent quantity are only available
   for very simple surface geometries. Any attempts at approximating
   complicated surfaces in terms of these simple geometries would
   necessarily be crude and therefore unprofitable.

4. Impact noise is a transient phenomenon. Consequently the associated
   energies in the frequency domain are distributed over a continuous
   band of frequenices. To calculate the radiated sound energy in
   each frequency band it would be necessary to weight the corresponding
   response of each impacted surface in the frequency domain with its
   radiation resistance in that frequency band. Clearly this is not a
   practical consideration.

At present much of the research on impact noise is concerned with
parts 1 and 2 of the above list under the general heading of "source
identification". Two methods of approaching this problem will be
presented here:

1. This first method called 'shielding' is a fairly crude but often
   effective means of identifying the sound radiating surfaces of a
   given machine. The method consists of first covering all the possible radiating surfaces with highly damped layered materials such
   as lead sheeting combined with a blanket of a porous sound absorber,
   say fibreglass. While monitoring the radiated sound field, the

layered materials are successively removed until an effective

radiating surface is identified.

2.   A second method requiring more sophisticated electronic equipment is

to identify the source by measuring the extent of correlation be-

tween a response point on the source and the sound pressure at

some field point.   The technique can be illustrated with the following

example.

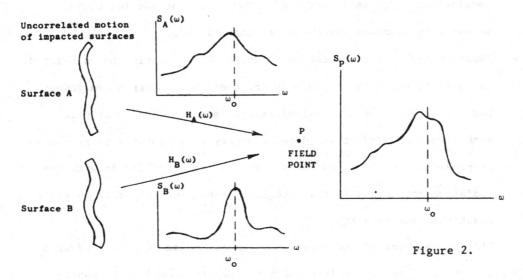

Figure 2.

The measured mean square spectral densities of the acceleration

response of surfaces A and B (see figure 2) indicate that both surfaces

contain energy at ω_o, the frequency of interest.   The mean square

spectral density of the pressure recorded at the field point indicates

that either one or both of the surfaces are radiating sound at the

frequency ω_o.   The problem is one of identifying which of the two sur-

faces is the more efficient source.

Before developing the details of the method, we must first introduce the ideas of cross-correlation and cross mean square spectral density. Recall that we defined the auto-correlation of the function $x(t)$ as $R_x(\tau) = E[x(t) \, x(t + \tau)]$. The corresponding Fourier transform of this quantity in the frequency domain was the mean square spectral density, $S(\omega)$ given by

$$S_x(\omega) = \frac{1}{2\pi} \int_{-\infty}^{\infty} R_x(\tau) \, e^{-i\omega\tau} \, d\tau \; . \tag{1}$$

We now introduce the cross correlation of two functions of time $x(t)$ and $y(t)$ and define

$$R_{xy}(\tau) = E[x(t)y(t+\tau)] \tag{2}$$

as the cross-correlation function. The corresponding transform in the frequency domain is the cross mean square spectral density given by

$$S_{xy}(\omega) = \frac{1}{2\pi} \int_{-\infty}^{\infty} R_{xy}(\tau) \, e^{-i\omega\tau} \, d\tau \, . \tag{3}$$

Returning to our problem, we can now <u>measure</u> the following additional information that we will need in our analysis:

$S_{pA}(\omega)$, $S_{pB}(\omega)$ —   cross mean square spectral density of the field pressure and the acceleration response of surface A and B respectively.

Next we assume that $S_p(\omega_o)$ can be related to $S_A(\omega_o)$ and $S_B(\omega_o)$ through the transfer functions $H_A(\omega_o)$ and $H_B(\omega_o)$ by

$$S_p(\omega_o) = |H_A(\omega_o)|^2 \, S_A(\omega_o) \tag{4}$$

and $S_p(\omega_o) = |H_B(\omega_o)|^2 \, S_B(\omega_o)$

It can be shown that the above transfer functions also relate the cross mean square spectral densities to $S_A(\omega_o)$ and $S_B(\omega_o)$ as

$$S_{pA}(\omega_o) = H_A(\omega_o) \, S_A(\omega_o)$$

$$S_{pB}(\omega_o) = H_B(\omega_o) \, S_B(\omega_o) \, . \tag{5}$$

By combining equations (4) and (5) we can eliminate the transfer function to arrive at a ratio of spectral densities which give us a measure of coherency between the source and the receiver. We define this ratio as the coherency function, $\gamma^2(\omega_o)$, which is given by

$$\gamma_A^2 = \frac{|S_{pA}(\omega_o)|^2}{S_p(\omega_o) S_A(\omega_o)}$$

$$\text{and} \quad \gamma_B^2 = \frac{|S_{pB}(\omega_o)|^2}{S_p(\omega_o) S_B(\omega_o)} \, . \tag{6}$$

The coherency function may vary between zero and unity and for our particular problem indicates to what extent the surfaces A and B contribute to the sound measured at the field point. Thus if a high value of γ_A^2 and a low value of γ_B^2 were obtained from the measured spectral densities, we could conclude that surface A is the more efficient of the two radiating surfaces.

Simple Model of Impacting Surfaces (reference 5.)

To gain an understanding of the parameters that influence the sound radiation from impacting surfaces, it is useful to consider the simple case of two spheres undergoing an elastic collision as shown in figure 3.

The acceleration A, of each sphere during the time d of impact is given approximately as

$$A = a_m \sin bt \qquad 0 \le t \le d \tag{7}$$

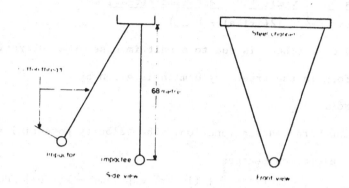

Figure 3   Schematic of sphere suspension apparatus

where a_m is the maximum value of the acceleration, and b is angular frequency of the half sine pulse given by π/d.

To examine the problem from a sound radiation point of view, we can begin by recalling the expression that we developed earlier for the acoustic velocity potential ϕ of a sphere of radius a, oscillating sinusoidally at a frequency ω and a velocity amplitude $\hat{v}_a$:

$$\phi = \frac{a^3 \hat{v}_a (1+ikr) \cos\theta \exp i[\omega t - \kappa(r-a)]}{r^2 [2(1+ika) - k^2 a^2]} \tag{8}$$

where r and θ locate the field point relative to the principle axis of the sphere. The velocity potential for any arbitrary velocity of the sphere (e.g. a transient function) can be obtained by using equation (8) to synthesize a solution in the frequency domain, provided a Fourier transform of the arbitrary velocity is available. Given a sphere with a velocity $v(\omega)$ along the principal axis, the corresponding velocity potential can be written as

$$\phi = \frac{a^3 \cos \theta}{2\pi r^2} \int_{-\infty}^{\infty} \frac{v(\omega)(1+ikr) \exp i[\omega t - k(r-a)] \, d\omega}{[2(1+ika) - k^2 a^2]} . \qquad (9)$$

If the velocity of a sphere is due to a unit impulse of acceleration,

then its transform in the frequency domain is given by

$$v(\omega) = \pi \delta(\omega) + \frac{i}{\omega} \qquad (10)$$

where $\delta(\omega)$ is the Dirac delta function. The velocity potential for a

unit impulse acceleration becomes

$$\phi_I = \frac{a^3 \cos \theta}{2\pi r^2} \int_{\infty}^{\infty} \frac{[\pi \delta(\omega) + \frac{i}{\omega}] (1+ikr) \exp i[\omega t - k(r-a)] \, d\omega}{[2(1+ika) - k^2 a^2]} \qquad (11)$$

If we now take the acceleration of the sphere given in equation (7) as

the forcing function, F(t) we can obtain the corresponding velocity po-

tential by convoluting ϕ_I with F(t) via the Duhamel integral:

$$\phi_F = \int_o^d \phi_I (t - \tau) F(\tau) \, d\tau \qquad (12)$$

where τ is the integration variable. The acoustic pressure produced by

the forcing function F is then

$$p = \rho_o \frac{\partial \phi_F}{\partial t} \qquad (13)$$

and thus the total acoustic pressure radiated by the impactor and the

impactee (see figure 4) is given by

$$p(r, \theta, t) = p(r_1, \theta_1, t) + p(r_2, \theta_2, t). \qquad (14)$$

Figure 5 shows a typical time history of the radiated sound pressure due

to the collision of two 12.7 mm spheres. Koss and Alfredson (reference

5), who formulated the above theory, have also demonstrated its excellent

agreement with experimental measurements. By examining the dependence

of the peak field pressure on the physical parameters of the colliding

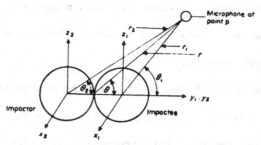

Figure 4   Coordinate system used to define positions in the
            radiation field.

Figure 5 Calculated sound
pulse for a 12.7 mm sphere
impacting a 12.7 mm sphere
for v_o = 1.52 m/s,
θ= 0.

o, wave from impactee
x, wave from impactor
o, total wave

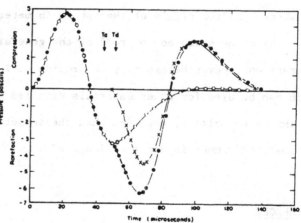

spheres, they have demonstrated the following parametric relations.

1. For a given sphere of radius a, the values of the first compression

    peak in the field pressure p vary with the velocity of impact v_o,

    according to the relation

    $$\frac{p}{E} = 1.29 \times 10^{-6} \; [\frac{r}{a}]^{-1.07} \; [\frac{v_o}{c}]^{-1.25} \qquad (15)$$

where E is modulus of elasticity, r is the radial distance from the im-

pact point, and c is the speed of sound. Initial observations that the

acoustic energy varies as v_o^2 have also been made on more complicated
structures but further research is necessary before any definite conclusions can be put forward.

2. The frequency at which the maximum energy occurs in the spectrum of
    the radiated sound is roughly proportional to ½d, where d is
    duration of the impact. This frequency, f, which is nearly inde-
    pendent of the impact velocities can be approximated by the formula

$$f = \frac{76.1}{a} \text{ Hz} \tag{16}$$

where a is the radius of the sphere in meters.

While there can be no proof of the general nature of the above
observations, nevertheless they do provide useful parametric relations
which can be used for other materials detailed shapes, hammer volumes,
and impact velocities, provided that the impact takes place at a point
and that the hammer is not too elongated.

REFERENCES

1. I. MALECKI 1969 Pergamon Press Physical Foundations of Technical
    Acoustics.

2. E. SKUDRZYK 1968 The Pennsylvania State University Press Simple
    and Complex Vibratory Systems.

3. E. MEYER and E. NEUMANN 1972 Academic Press Physical and Applied
    Acoustics (originally published in German under the title
    Physikalische und Technische Akustik 1967 F. Vieweg & Sohn GmbH).

4. L.L. KOSS and R.J. ALFREDSON 1974 Journal of Sound and Vibration
    34(1) 11-33 Identification of transient sound sources on a punch
    press.

5. R. J. ALFREDSON and L.L. KOSS 1973 Journal of Sound and Vibration
    27 59-75 Transient sound radiated by spheres undergoing an elastic
    collision.

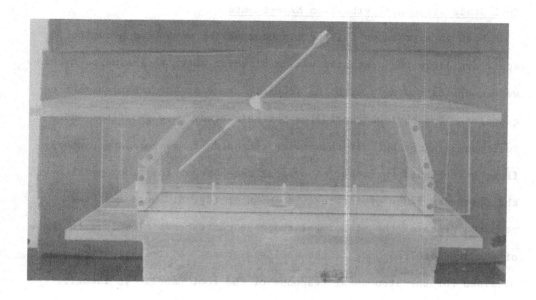

Figure 9   1:12 scale model of van

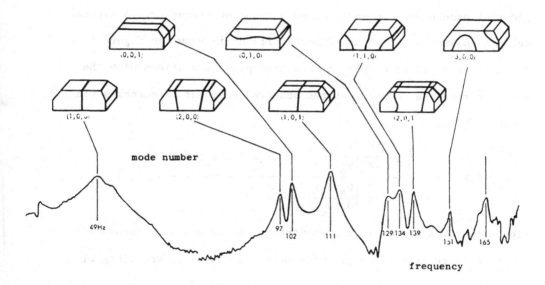

Figure 10.

Full Scale Structural Vibration Experiments

With an expression for the response of the uncoupled acoustic
space available the next step was to discover which enclosing surfaces
were coupling strongly with the acoustic space at the troublesome frequencies. Two obvious candidates for this study were the floor and
roof panels since both of these were composed of several large, nearly
flat, unstiffened surfaces. A typical response of the floor panel over
the frequency range of interest is shown in fig. 8. Note that the
resonant peaks of the structural response corresponds closely with those
of the acoustic response of the enclosed space. A similar curve was
obtained for the structural response of the roof panel. To identify
which of the vibrating surfaces coupled most strongly with the acoustic
modes of the space, the velocity distributions $\dot{v}(x_o, y_o, z_o)$ of the floor
and roof panels were measured at each resonant frequency. A typical
velocity distribution of the floor at 113 Hz is shown in figure 11.
The next step was to couple the measured panel vibrations with the
acoustic response function at the observed resonant frequencies via
equation (24), i.e.

$$J.A.F. = \frac{\oint [\frac{P_o(x,y,z)}{Q_o(x_o,y_o,z_o)}] \; v \; (x_o, \; y_o, \; z_o) \; dS}{\oint \left| \frac{P_o(x,y,z)}{Q_o(x_o,y_o,z_o)} \right| \; dS} \; .$$

Figure 11 gives a graphical presentation of the acoustic response
function in terms of a single mode over the floor surface along with
the vibration response function of the floor panel at the same frequency.
Combining these two response functions in the above equation gives a

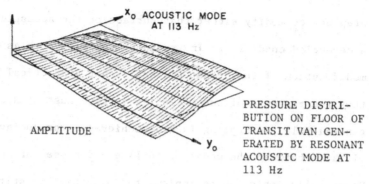

PRESSURE DISTRI-
BUTION ON FLOOR OF
TRANSIT VAN GEN-
ERATED BY RESONANT
ACOUSTIC MODE AT
113 Hz

Figure 11.

Coupling of Modal
Geometry

Joint Acceptance
Function = 0.83

GEOMETRY OF FLOOR
PANEL RESONANT
RESPONSE AT 113 Hz

joint acceptance function of 0.83. By performing similar operations on

the remaining acoustic and structural response functions, the most

strongly coupled surfaces were identified for vibration control treat-

ment.

Recommendations

   With identification of the strongly coupled surfaces, the next step was to modify either that surface or the acoustic space to produce

a decoupled condition.   In transport vehicles such as a transit van,

modification of the acoustic space is not a practical consideration and

thus the response of the enclosing surface must be altered.   As a

general rule, decoupling is best achieved by producing a geometrical

mismatch between the modes of surface and those of the enclosed space.

Practically, this can be implemented by attaching stiffeners to the

enclosing surfaces to produce a modal response where the antinodal

regions of the surface coincide with the nodal regions of the acoustic

space and vice versa.

References

1.   M, PETYT, J. LEA and G.H. KOOPMANN 1975 Journal of Sound and
     Vibration 45(4) 495-502 A finite element method for determining
     the acoustic modes of irregular shaped cavities.

2.   G.H. KOOPMANN and H.F. POLLARD 1976 Journal of Sound and Vibration
     46(2) 302-305 A joint acceptance function for structural-acoustical
     coupling problems.

3.   R.J. DAMKEVALA, J.E. MANNING and R.H. LYON 1974 US Department of
     Transportation Report No. DOT.TSC.OST. 74.5 Noise control handbook
     for diesel-powered vehicles.

4,   Morse, P.M., 1948 Vibration and Sound, New York, McGraw Hill.

INTERIOR NOISE IN TRANSPORTATION VEHICLES

This lecture treats the subject of interior noise in transport vehicles such as observed in the cabs of trucks or tractors or in the passenger compartment of an automobile. Generally, the noise problem is a low frequency one where the acoustic modes of the enclosed space are excited into resonance by the vibration of the enclosing surface. In this low frequency range ($\sim$50-300 Hz) the acoustic modes are governed mainly by the geometry and dimension of the enclosed space and thus their analysis is fairly straightforward.

Analysis of the forced response of the enclosing surfaces, however, is exceedingly difficult. Firstly, the enclosing structure is composed of several smaller substructures ranging from doubly curved laminated glass windows to damped, stiffened sheet metal panels. If a 'modal' response is apparent, either in the overall structure or individual sub-structure, it is difficult to identify which surfaces couple most efficiently with the corresponding acoustic modes of the enclosed space. A second problem is the identification of the input forces. The forcing of the structural 'modes' can occur through many different mechanisms, many of which are interactive. Examples of force inputs to the enclosing structural surfaces on a passenger automobile are as follows:

1) Vibrations of the engine/power train which are transmitted through their corresponding suspension systems.

2) Road roughness transmitted through the wheel/spring suspension systems.

3) Acoustic excitation due to airborne noise from intake and exhaust ports, cooling fans, etc.

4) Aerodynamic forces due to the unsteady flow around the exterior surfaces of the vehicle.

In spite of all the complications mentioned above, it is possible to analyse the complex system by first treating the acoustic space as an uncoupled system wherein the enclosing boundaries are replaced by rigid boundaries of the same geometry. With the acoustic response of the uncoupled system available, it is then possible to identify structural modes on the actual surface which may or may not couple strongly with the corresponding acoustic modes.

Acoustic Modes

To illustrate the technique of modelling the response of the uncoupled acoustic space, we first want to develop the concepts of acoustic modes and overall acoustic response a little further. To begin, let us

assume that our acoustic space is a rectangular shape of dimensions

L_x, L_y and L_z (see figure 1) where $L_x > L_y > L_z$. Further we will assume

that the enclosing surfaces of the space are acoustically 'hard', i.e.

the normal component of the acoustic perturbation velocity vanishes at

the surface.

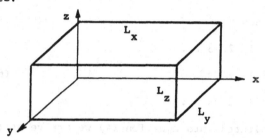

Figure 1.

Beginning with the three dimensional wave equation in cartesian

coordinates in terms of acoustic pressure, p,

$$\frac{\partial^2 p}{\partial x^2} + \frac{\partial^2 p}{\partial y^2} + \frac{\partial^2 p}{\partial z^2} = \frac{1}{c^2} \frac{\partial^2 p}{\partial t^2} \qquad (1)$$

we can write a harmonic time dependent solution of the form

$$p = \hat{p} \, \psi(x) \, \psi(y) \, \psi(z) \, \sin \omega t \qquad (2)$$

where $\psi(x)$, $\psi(y)$ and $\psi(z)$ are the shape functions of the acoustic modes

in the x, y and z direction respectively, and ω is the angular frequency.

For the 'hard wall' boundary condition considered in this model, the

shape functions are cosines and thus

$$p = \hat{p} \, \cos k_x x \, \cos k_y y \, \cos k_z z \, \sin \omega t \qquad (3)$$

where k_x, k_y, and k_z are the components of the wave number $k = \omega/c$.

Twice differentiating this solution with respect to the time and

space variables and substituting the results into equation (1) gives a relation between the wave numbers as

$$k^2 = k_x^2 + k_y^2 + k_z^2 = (\frac{2\pi f}{c})^2 \tag{4}$$

From equation (3) it can be seen that the pressure is maximum at the

boundary planes $x = 0, L_x$; $y = 0, L_y$; and $z = 0, L_z$. Consequently,

$$|\cos k_x L_x \cos k_y L_y \cos k_z L_z| = 1 \tag{5}$$

To satisfy this condition, the arguments of the cosine functions must

take on the following values:

$$k_x L_x = n_x \pi \qquad n_x = 1, 2, 3 \ldots$$
$$k_y L_y = n_y \pi \qquad n_y = 1, 2, 3 \ldots \tag{6}$$
$$k_z L_z = n_z \pi \qquad n_z = 1, 2, 3 \ldots$$

If these values are substituted into equation (4) we arrive at the

expression for the characteristic frequencies of the enclosed acoustic

space:

$$f_{n_x, n_y, n_z} = \frac{c}{2} [(\frac{n_x}{L_x})^2 + (\frac{n_y}{L_y})^2 + (\frac{n_z}{L_z})^2]^{\frac{1}{2}}. \tag{7}$$

Each trio of numbers (n_x, n_y, n_z) corresponds to a 'normal mode' of the

acoustic space for which the spatial distribution of the sound pressure

can be calculated from equation (3) and the frequency is given by

equation (7). The series of numbers gives the number of nodal planes

parallel to the corresponding coordinate axis. For example, the (1,0,1)

mode is shown in figure 2.

$$n_x = 1$$
$$n_y = 0$$
$$n_z = 1$$

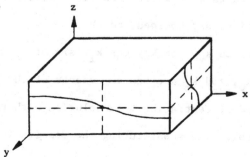

Figure 2.

Modal Response due to a Simple Source

With the individual modes of the acoustic space described in the frequency and spatial domain, our next step is to add them all together to obtain an expression for the overall response of the space to a given source excitation. Synthesizing the forced response of the acoustic space in terms of undamped modes raises a problem, however, since the response at each modal frequency would be unbounded. The modal can be made more realistic by introducing damping in the form of a complex frequency term in the wave equation, the real part of which represents the resistive part of the impedance assigned to each individual standing wave (see reference 4). This step is justified if the values chosen for the damping are small enough such that the undamped shape functions remain as valid first order approximations to the modified wave equation. Summing over all of the lightly damped modes, N, the pressure p at a field point (x, y, z) within an acoustic space of volume V, due to a simple source of strength Q_o at a source point (x_o, y_o, z_o) oscillating at an angular frequency ω, is given by

$$P_o(x,y,z,\omega) \doteq \frac{\rho_o c_o^2 Q_o \exp i\omega t}{V} \sum_N \frac{\psi_N(x,y,z)\ \psi_N(x_o,y_o,z_o)}{\Lambda_N\ [\frac{2\omega_N \kappa_N}{\omega} + i(\frac{\omega_N^2}{\omega} - \omega)]}, \quad (8)$$

For the Nth mode, $\psi_N(x,y,z)$ is the shape function, ω_N is the resonant frequency, κ_N is the assigned damping ratio, and Λ_N is the norm of the shape function. In the example being considered, ω_N is given by equation (7), $\Lambda_N = \epsilon_{n_x} \epsilon_{n_y} \epsilon_{n_z}$ where $\epsilon_o = 1$ and $\epsilon_2 = \epsilon_3 = \ldots = 2$, and

$$\psi_N(x,y,z) = \cos \frac{n_x \pi x}{L_x} \cos \frac{n_y \pi y}{L_y} \cos \frac{n_z \pi z}{L_z}. \quad (9)$$

Note that each normal mode has an amplitude which is 1) proportional to the product of the shape functions evaluated at the field and source points, and 2) proportional to its impedance given by

$$z_N = [\frac{2\omega_N \kappa_N}{\omega} + i\,(\frac{\omega_N^2}{\omega} - \omega)\,]^{-1}. \tag{10}$$

The contribution from each mode to the overall pressure field is thus dependent on the following:

1) The position of the source in the enclosed space.

In the example considered, a source positioned near the geometric centre of the rectangular space would only excite modes with even numbers of nodal planes while a source placed near a corner would excite all modes equally.

2) The position of the receiver in the enclosed space.

The dependency of the receiver pressure field on position is identical to that of the source since both source and receiver share the same shape functions.

3) The amount of acoustic damping in each mode.

Modal Response due to Distributed Sources

In the previous sections we examined the response of an acoustic space due to a simple source. We now want to extend the analysis to include the effects of distributed sources on the boundary to illustrate how the acoustic modes of a space couple with the out-of-plane vibration of the enclosing surfaces. In terms of equivalent sources, this motion can be represented with a continuous distribution of simple sources which has a velocity distribution identical to that of the surface itself. Thus, for an enclosing surface S_o, vibrating at an angular frequency ω,

with an out-of-plane velocity distribution $v(x_o, y_o, z_o)$, the pressure
at a field point (receiver) would be given by a modified form of equation
(8) as

$$p(x,y,z) \doteq \frac{\rho_o c_o^2 \exp i\omega t}{V}$$

$$\sum_N \frac{\psi_N(x,y,z) \int_s v(x_o,y_o,z_o) \ \psi_N(x_o,y_o,z_o) \ dS_o}{\Lambda_N [\frac{2\omega_N \kappa_N}{\omega} + i \ (\frac{\omega_N^2}{\omega} - \omega) \ ]} \qquad (11)$$

It should be noted that, although the enclosing surfaces undergo vibration, they still must reflect incident pressures as 'acoustically
hard' surfaces when considered from an acoustic point of view.

To illustrate how equation (11) may be used to identify which enclosing surfaces of a space couple most efficiently with a given acoustic
mode, consider the following problem. A rectangular box of dimensions
L_x, L_y and L_z (see figure 3) is composed in such a way that the floor
and ceiling surfaces undergo uncorrelated, out-of-plane vibrations
while the other enclosing surface remain rigid. The dimensions of the
box are such that the frequency ω_{100} of the first 'hard wall' acoustic
mode (1,0,0) is well separated from the next higher modes. The floor
and ceiling surfaces both vibrate at the frequency $\omega = \omega_{100}$ with the
following velocity distributions:

Ceiling $\quad v_C (x_o, y_o) = \hat{v}_C \sin \frac{4\pi x_o}{L_x} \quad$ at $\quad z = L_z \qquad (12)$

Floor $\quad v_F (x_o, y_o) = \hat{v}_F \sin \frac{2\pi x_o}{L_x} \quad$ at $\quad z = 0 \qquad (13)$

The corresponding shape function for the (1,0,0) acoustic mode is given
by

$$\psi_{100} (x_o, y_o) = \cos \frac{\pi x_o}{L_x} \qquad (14)$$

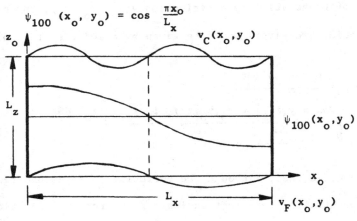

Figure 3.

The above equations can be substituted into equation (11) to determine
which of the two vibrating surface couples more strongly with the
corresponding acoustic mode. Considering the ceiling vibration first,
the field pressure for the (1,0,0) acoustic mode generated at any point
in the box is given by

$$p_C(x,) = \frac{\rho_o c_o^2 \exp i\omega_N t}{V\,(2\,\kappa_N)} \cos \frac{\pi x}{L_x} \int_{L_x}\int_{L_y} \hat{v}_C \sin \frac{4\pi x_o}{L_x} \cos \frac{\pi x_o}{L_x} dx_o\, dy_o$$

$$= \frac{\rho_o c_o^2 \exp i\omega_N t}{V\,(2\,\kappa_N)} \cos \frac{\pi x}{L_x} \left[\frac{8}{15\pi} \hat{v}_C L_x L_y\right]. \qquad (16)$$

In a similar manner, the field pressure generated by the floor vibration
is found to be

$$p_F(x) = \frac{\rho_o c_o^2 \exp i\omega_N t}{V(2\,\kappa_N)} \cos \frac{\pi x}{L_x} \left[\frac{4}{3} \frac{\hat{v}_F}{\pi} L_x L_y\right]. \qquad (17)$$

If the two surfaces are vibrating with the same amplitude, the ratio of
the two pressures generated by them in logarithmic form reduces to

$$20 \log_{10} \frac{p_F}{p_C} = 20 \log_{10} 2.5 \doteq 8 \text{ dB}. \qquad (18)$$

The pressure field produced by the floor vibration is 8 dB higher than that produced by the ceiling vibration. Thus when considering which surface to treat from a noise control point of view, reduction of the floor surface response would produce the greater reduction in interior sound pressure level.

Interior Noise Control

When considering methods of controlling noise levels in transport vehicles equation (11) can be used to provide useful engineering guidelines. Considering each term separately, the following noise control measures can be adopted.

1) The enclosing surface velocities $v(x_o, y_o, z_o)$, should be held to a minimum. This can be achieved with the following engineering designs:

   a) Barrier type materials should be used as enclosing surfaces exposed to noise from the engine/transmission compartments. When limp materials cannot be used, damping may have to be bonded to the enclosing surfaces directly to reduce their vibration levels. For floors, heavy duty floor mats or thick rubber boots may be employed. In extreme cases, a layer of lead loaded vinyl may be inserted under the mats.

   b) The enclosing surfaces should be vibration isolated from the engine and main chassis. Where this is not possible, as in unit body construction, rubber isolator pads on mounting locations of the power train should be employed.

2) The acoustic damping κ_N in each mode should be made as high as possible by fitting all exposed surfaces with sound absorbing materials,

especially the roof, side walls and doors.

3) In the design stage of choosing a shape for the enclosing space,
flat, reflecting surfaces which are opposite to one another should be
slanted rather than arranged parallel to one another. Although acoustic
modes are still set up in irregular shaped spaces, the anti-nodal
positions of the modes are more evenly distributed than those which
occur in regularly shaped spaces.

Joint Acceptance Function

In the previous section, we combined a single vibratory mode of
a plate with a single acoustic mode of an enclosed space to predict the
extent of their mutual coupling in terms of the interior acoustic
pressure. We now want to extend that analysis to include several acoustic modes in the overall acoustic response. We will also introduce a
normalising procedure that will quantify the coupling between an arbitrary surface vibration and the acoustic modes of the enclosed space in
terms of a single quantity called the joint acceptance function.

To begin, we can write equation (8) as a ratio of the field
pressure $p_o(x,y,z)$, to the unit volume velocity of the point source
$Q_o(x_o, y_o, z_o)$, given by

$$\frac{p_o(x,y,z)}{Q_o(x_o,y_o,z_o)} \doteq \frac{\rho c^2 \exp i\omega t}{V} \sum \frac{\psi_N(x,y,z)\ \psi_N(x_o,y_o,z_o)}{\Lambda_N\ z_N^{-1}} . \quad (19)$$

Multiplying each side of equation (19) with a surface normal velocity
distribution $v(x_o,y_o,z_o)$ and integrating over the entire surface area
S_o gives, after examination of equation (11)

$$p(x,y,z) = \int_S \left[\frac{p_o(x,y,z)}{Q_o(x_o,y_o,z_o)}\right]\ v(x_o,\ y_o,\ z_o)\ dS_o . \quad (20)$$

Thus, if the shape functions and the corresponding impedances of the acoustic modes are available for a given space enclosed by nearly 'acoustically hard' surfaces the acoustic response $[P_o/Q_o]$ can be calculated by summing over the modes of interest. In turn, this response can be used in equation (20) to determine the interior pressure field $p(x,y,z)$ produced by the motion of the enclosing surfaces vibrating sinusoidally with an arbitrary velocity distribution $v(x_o, y_o, z_o)$.

The use of equation (20) to quantify coupling, however, is limited somewhat by its functional dependence on frequency and field position. At first glance, it appears that the integral would have to be evaluated as a function of frequency within the frequency range of interest. In practical terms, however, this is unnecessary since the range of frequencies is marked only by a few selected structural and/or acoustical resonances and thus, the integral would only have to be evaluated at a small set of discrete frequencies within a specified range. The problem of field position dependency can be treated in the following manner. A distribution of surface velocities can be arbitrarily chosen in equation (20) which produces a maximum pressure at a given field point. Such a distribution is first found by writing the complex quantity $[P_o/Q_o]$ in terms of its magnitude and phase, i.e.

$$| P_o/Q_o | \ \exp (i\theta_{P_o Q_o}) \qquad\qquad (21)$$

where $\theta_{P_o Q_o}$ is the phase angle between the field pressure and the unit volume velocity source. If the arbitrary velocity distribution normal to the surface element dS_o is specified as

$$v(x_o, y_o, z_o) \ = \ |Q_o| \ \exp(- i \ \theta_{P_o Q_o}) \qquad\qquad (22)$$

the corresponding pressure at the given field point would assume a

maximum value, $p_{max}(x,y,z)$ given as

$$p_{max}(x,y,z) = \int_{S_o} \left| \frac{P_o}{Q_o} \right| \, dS_o \, . \tag{23}$$

$p_{max}(x,y,z)$ would be the total pressure at a given field point obtained

by summing over all the unit volume velocities distributed over the

surface S_o. This is a convenient quantity on which to normalise since

the ratio of the field pressure $p(x,y,z)$ due to any arbitrary velocity

distribution $v(x_o,y_o,z_o)$ to $p_{max}(x,y,z)$ gives a relative measure of the

structural/acoustic coupling. The ratio $p(x,y,z)$ to $p_{max}(x,y,z)$ is

referred to as the joint acceptance function, J, where:

$$J = \frac{\int_{S_o} \left[\frac{P_o(x,y,z)}{Q_o(x_o,y_o,z_o)} \right] v(x_o,y_o,z_o) \, dS_o}{\int_{S_o} \left| \frac{P_o(x,y,z)}{Q_o(x_o,y_o,z_o)} \right| \, dS_o} \, . \tag{24}$$

The expression 'joint acceptance' was originally introduced in the

analysis of the response of structures to random acoustic loading and

gives a measure of the extent to which a particular structural mode

accepts a given acoustic trace wavelength. In this work, it can be

interpreted in a similar manner, i.e. in terms of modal analysis. Thus

for any vibrating surface enclosing an acoustic cavity, the joint accep-

tance function defines the extent to which the acoustic modes of a space

geometrically 'accept' the structural modes of an enclosing surface at

a given frequency.

Application of the Joint Acceptance Function

      The usefulness of the joint acceptance function can be illustrated with the following simple example. A rectangular enclosure of scaled dimensions $L_x = 0.236$ m, $L_y = 0.128$ m, and $L_z = 0.112$ m is constructed of surfaces which provide light damping in the first few acoustic modes as shown in figure 4. The bottom enclosing surface ($z = 0$) vibrates as a plate with the velocity functions:

$$v_{mn}(x_o, y_o) = \hat{v}_{mn} \sin \frac{m\pi x_o}{L_x} \sin \frac{n\pi y_o}{L_y}. \qquad (25)$$

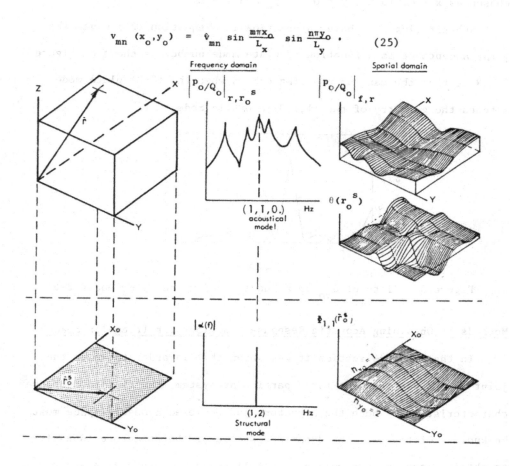

Figure 4. Coupling of Hard Walled Acoustic Cavity with a Flexible Bottom

We want to examine the extent of structural/acoustic coupling at the frequency of the (1,1,0) acoustic mode for various plate mode shapes. To apply equation (19) we proceed by summing over the acoustic modes at and around the (1,1,0) mode frequency. For this particular case, modes 2-6 cover an adequate range of frequenices. To minimise the field position dependency in equation (20) a field point near a corner is chosen as $x = 0.915 L_x$, $y = 0.156 L_y$ and $z = L_z$.

Substituting the above information into equation (24) gives the joint acceptance as a function of plate mode number as shown on figure 5. Note that the maximum coupling occurs when the (2,2) plate mode matches the geometry of the (1,1,0) acoustic mode.

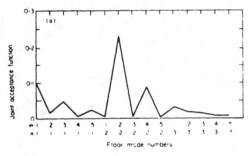

Figure 5.   Plots of J_{mn} as a function of m and n for modes 2-6

Methods of Obtaining Acoustic Response Functions for Irregular Spaces

In the previous section it was shown that in order to write the joint acceptance function for a particular system, the dynamic response characteristics of both the structure and enclosed acoustic space must be known. Of the two, the dynamic response of the structure is easier to obtain since measurements on the full scale prototype give exactly the information required in equation (24), i.e. the coupled frequencies and corresponding modal geometries. The measurement of the acoustic response function is complicated beyond practicality by the coupled motion of the enclosing structure and thus it becomes necessary to construct, either physically or mathematically, an equivalent model of the uncoupled acoustic space. Two modelling techniques which can be used to generate acoustic response functions for irregularly shaped enclosures are described as follows.

One method of acquiring the acoustic response of an irregularly shaped space is to construct a physical model of the space to a suitable scale. The response function $[P_0(x,y,z)/Q_0(x_o,y_o,z_o)]$ can be obtained by exciting the space with a small moveable acoustic source located in the plane of the enclosing surfaces (see figure 6) and measuring the corresponding acoustic pressure at a fixed field point.

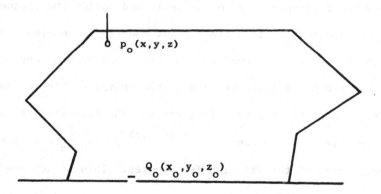

Figure 6.

Before going into the details of the method, however, a few points on acoustic modelling techniques should be mentioned.

a) The walls of the model should be constructed of a material which is rigid enough to prevent them from responding to the interior acoustic

pressure fields.

b) The acoustic source should approximate to a point source and thus

    should have dimensions which are small compared to an acoustic

    wavelength.

c) The temperature should be monitored carefully throughout each test

    to ensure consistancy of results acquired over long periods of time.

d) The overall shape and dimension of the model should conform to that

    of the full scale prototype. However, details such as rounded

    corners, etc. with dimensions that are small compared to a typical

    wavelength of sound have minimal influence on the shape functions of

    the space and can be approximated with simpler geometries.

With a suitable model and supporting apparatus available, the modal

frequencies of the space can be determined by driving the acoustic

source through the frequency range of interest and noting the frequencies

at which the pressure at the fixed field point undergoes a maximum res-

ponse. (During this step care must be taken to avoid placing the source

or receiver near the nodal point of an acoustic mode). Following iden-

tification of the acoustic resonant frequencies, the next step is to

generate an acoustic response function $[P_o{}^{(x,y,z)} /Q_o(x_o,y_o,z_o)]$ for a

given surface at each of the resonant frequencies. This is achieved by

recording the amplitude and phase of the pressure at a fixed field

point while moving the source over those surfaces for which corresponding

full scale structural response data are available. Finally. at each

resonant frequency, the acoustic response function can be combined with

the structural response function via equation (24) to quantify the degree of structural/acoustic coupling.

A second means of acquiring an acoustic response function for an irregularly shaped enclosure is the acoustic finite element method which necessitates the use of a fairly elaborate computer program. In this mathematical method, the volume of an irregularly shaped space is substituted with a set of smaller, simpler, shaped volumes which, in combination, approximate to the original shape. By choosing appropriate shape functions which satisfy the pressure conditions on the boundaries of these smaller volumes, an acoustic response function in terms of the overall shape functions of the original volume can be constructed. An example of this modelling technique is illustrated in figure 7, where the acoustic space of a transit van is idealised in terms of sixteen finite elements. The corresponding frequencies and shape functions of the first few acoustic modes are also shown (scale 1:12). Note that the slanting of the reflecting surfaces results in shape functions that are characterised by curved rather than flat nodal planes.

Case Study

This section presents a few of the results of an investigation on the low frequency interior noise of a lightweight transit van and illustrates how the joint acceptance function can be used as a guide to reducing the interior noise level.

Full Scale Acoustic Measurements

During the measurement of the sound pressure level in the interior of the van, it was observed that several strong resonant frequencies occurred in the frequency spectrum as the engine speed was slowly in-

creased from idle to normal operating conditions (approx. 1000-5000 rpm).

The peak frequencies along with their amplitudes as measured at the ear

of a front seat passenger are given in figure 8. (Note that the pres-

sure are given in dB lin).

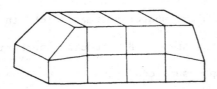

Figure 7(a)   Idealisation of half van model

Acoustic frequencies of model van
(a) Symmetric modes

Mode	Experimental (Hz)	Theoretical (Hz)
1	606	593
2	1174	1150
3	1549	1556
4	1613	1605
5	1817	1829
6	1992	2026

(b) Antisymmetric modes

Mode	Experimental (Hz)	Theoretical (Hz)
1	1220	1168
2	1352	1317
3	1675	1634
4	1996	1956
5	2021	1997
6	2176	2187

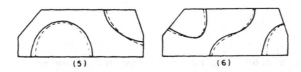

(1)   (2)

(3)   (4)

(5)   (6)

--- Experimental
___ Theoretical

Figure 7(b) Node lines for the acoustic modes of a model van

Model Experiments

To identify the acoustic shape functions which corresponded to the

frequencies of the pressure peaks observed in the full scale experiments,

a 1:12 scale model of the van was constructed. (see figure 9).

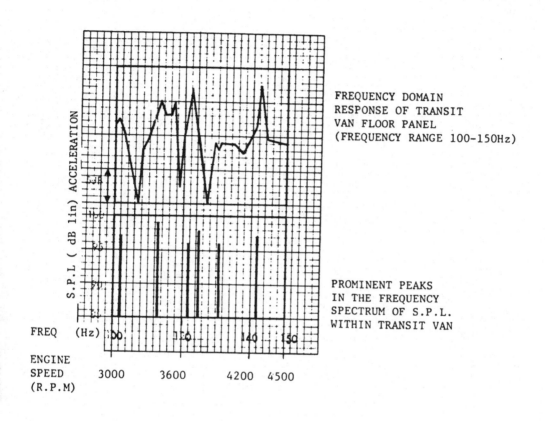

FREQUENCY DOMAIN
RESPONSE OF TRANSIT
VAN FLOOR PANEL
(FREQUENCY RANGE 100-150Hz)

PROMINENT PEAKS
IN THE FREQUENCY
SPECTRUM OF S.P.L.
WITHIN TRANSIT VAN

Figure 8

Experiments on the model verified the existence of the observed acoustic

modes and produced a measure of their corresponding shape functions

expressed in terms of pressure distributions. The results of these

model studies are shown in figure 10. With the above modal information

available, it was thus possible to generate the required response

function $P_o(x,y,z)/Q_o(x_o,y_o,z_o)$ over those enclosing surfaces for which

coupling was suspected.

FLOW NOISE
(Notes for Lectures 16/17; 21-28)
SOURCES OF FLOW ASSOCIATED NOISE

P.O.A.L. DAVIES
Professor of Experimental Fluid Dynamics
Institute of Sound and Vibration Research
The University, Southampton SO9 5NH,
ENGLAND.

1.0   Introduction

Many noise sources that represent a major environmental noise
nuisance, or a hazard to health in industry, are of aerodynamic origin.
That is, they are associated with relatively high speed flows. One can
classify such sources by the manner in which they radiate noise to the
acoustic far field. (see A1.2)

Alternatively, aerodynamic sources are often popularly classified by
the flow system associated with them, for example; jet noise, fan noise
vortex shedding noise and so on. Finally they may be classified by the
dominant flow source mechanisms associated with them, for example;
mixing noise, edge tones, combustion noise, cavitation noise etc.

Sources of first order - or acoustic monopoles - radiated sound
uniformly in all directions. They can be treated theoretically as a
consequence of an unsteady introduction of mass at a point. In practice
such sources must be acoustically compact, that is their physical
dimensions should be small compared with the acoustic wavelength.
Practical examples of such sources include acoustic radiation by a siren,
a Hartmann generator, the pulsating flow at an internal combustion engine
or reciprocating compressor intake or exhaust pipe, the 'thickness' noise
of fast moving blade tips or bodies and explosions or rapid combustion
processes. Continuous sources provide discrete tones in the radiated
acoustic noise field.

Sources of second order – or acoustic dipoles – have a spatial
radiation pattern which varies with direction according to a cosine law.
They can be analysed in terms of a fluctuating point force, the direction
of the force vector providing the reference direction for the radiation
pattern.   Practical examples include unsteady lift and drag forces on
moving blades or bodies, unsteady forces produced at edges or on surfaces
by impinging flows or periodic vibrations of wires or other bodies immersed
in a flow.   The singing of telephone wires in strong winds represents
one such naturally excited example, the vibration being sustained by
vortices periodically shed from the wire.   Continuous sources provide
both discrete tones and broad band acoustic radiation in the far field.

Sources of higher order – or quadrupoles – radiate sound with a dis-
tribution which is described by a $\cos^2\theta$ or $\cos\theta\sin\theta$ relationship.   Mixing
noise is associated with the unsteady momentum fluctuations in jets or
wakes.   However, such sources are usually associated in practical cases
with sources of lower (and higher) order.   Mixing noise is broad band in
character, though the intensity will vary with frequency, rising to a peak
and then decaying slowly as frequency increases.

Classification of sources by flow system or source mechanism is an
alternative approach.   But complications can arise when making a quantitative prediction of the noise climate produced, in a particular practical
case, since normally several noise producing mechanisms are acting together.
This means that sources of different order are present simultaneously.

1.1   Specification of source strength and source characteristics

Sound is a form of mechanical energy.   It is therefore appropriate
to describe the generation process by rate of energy production, or the
sound power generated by the source.   Of importance also is the rate
energy passes through unit area of space.   This provides a measure of
energy concentration and is known as the sound intensity.   When considering sound sources one is generally concerned with the specification of
the following characteristics.

(a)   Sound pressure level

The sound pressure level is based on the root mean square pressure
$\sqrt{p^2}$ measured at same point (i.e. the standard deviation of fluctuating
pressure σ_p).   It is normally expressed as

$$SPL = 20 \log \sqrt{p^2}/\sqrt{p^2}_{ref}$$

where $\sqrt{p^2}_{ref}$ is a base pressure of 20 micropascals $(2\times10^{-5} Nm^{-2})$.
Measurements are normally made with a microphone connected to electronic
equipment to calculate the appropriate mean square time averages.   Since

the measurements are made in electrical signals, sound pressure level is often loosely termed sound power, which is misleading.

The acoustic power is the integration, over a surface surrounding the source, of the acoustic energy flux through the surface (intensity). For a point source radiating spherically the sound power is given by

$$W = \frac{4\pi}{\rho_o c_o} \, \overline{p^2} \; .$$

Sound level meters may be scaled to read decibels ref. 10^{-12} Watts, i.e. scaled in "sound power level" or P.W.L. All meters normally record root mean square pressure $\sqrt{\overline{p^2}}$, so the appropriate factor – of about 400 – is included in the scaling. Thus care should be exercised to examine carefully the way in which a measurement or instrument is scaled. Scaling in sound power level probably includes a tacit assumption that the recorded pressure arises from a plane acoustic wave so that the meter reading represents power per unit area or $\overline{p^2}/\rho_o c_o$, i.e. the intensity. For the relationships in specific circumstances see (Table 3).

(b)   Acoustic efficiency

This is the ratio of the acoustic power generated by a source to that mechanical chemical or electrical power which is supplied to it. It normally lies between 10% for a siren to 0.001% for flow mixing noise.

(c)   Frequency characteristics

Many sources emit sound at a number of discrete frequencies (tones) others emit sound over a continuous range of frequencies (broad band noise). The frequency distribution of acoustic energy is described by a frequency spectrum. Measurements of spectral distribution can be presented in several ways, the commonest being octave band, 1/3 octave band, or narrow band analysis. The standard centre band frequencies are listed in Table 1. The octave band centre frequencies are indicated by asterisks.

The bandwidth lies between 70% and 26% of spread from the centre frequencies of each band. The mean square pressures associated with each band are added to give the total contribution to that band, The sound pressure level in the 1 Hz band is termed the spectrum level or spectral density. The relationship between the spectrum level, band level and bandwidth Δf is

Spectrum level = SPL_{band} - 10 log Δf dB.

Spectral measurements are expressed as sound pressure level, in dB, and
regarded as sound power, reference 10^{-12} Watts.   Since the energy is
integrated over the band, the general shape and appearance of the
different type analyses will differ, as illustrated in the figure below.

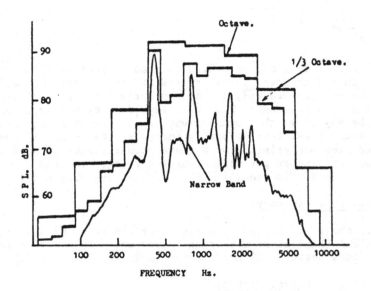

FIG. 1   COMPARISON OF OCTAVE, 1/3 OCTAVE AND
          NARROW BAND SPECTRA OF THE SAME SOUND.

(d)   Practical measurements of sound power

     Because the ear is not equally sensitive to all audio frequencies,
the overall mean sound pressure level sometimes called the overall SPL
and recorded as dB (lin), does not correlate well with subjective response
to noise, in terms of annoyance or loudness.   It is therefore common
practice to take account of the variation of sensitivy of the ear, by
weighting the signal either before or after analysis.   The A-weighting
is most generally applied and is set out in Table II.   The resulting
measure is termed db(A), which represents a frequency weighted sound
pressure level.

     Alternative methods of assessing noise are provided by the Noise

Rating (NR) and Noise Criterion (NC) curves.   NC curves are primarily
employed for architectural spaces while NR are more generally accepted
as industrial noise design or rating criteria.   The use of NR curves is
described in ISO R/1996 (1970).

(e)   Directivity

We have already noted that sources of different order exhibit
differing directional sound radiation patterns.   Many practical aerodynamic sources are complex combinations of sources of different order
and may exhibit complex directional sound radiation patterns.   Since
their effective scale is then small, compared with the acoustic wavelength,
compact sources will radiate more uniformly at relatively low frequencies
than at relatively high frequencies, when source dimensions are no longer
a small fraction of the acoustic wavelength.

TABLE 1

1/3 Octave Band Centre Frequencies (Hz)

25	*250	2500
* 31.5	315	3150
40	400	* 4000
50	*500	5000
* 63	630	6300
80	800	* 8000
100	* 1000	10000
*125	1250	12500
160	1600	* 16000
200	* 2000	20000

TABLE I

dB(A) Weighting, Relative Response

Frequency (Hz)	Weighting (dB)	Frequency (Hz)	Weighting (dB
20	-50.5	500	-3.2
25	-44.7	630	-1.9
31.5	-39.4	800	-0.8
40	-34.6	1000	0
50	-30.2	1250	+0.6
63	-26.2	1600	+1.0
80	-22.5	2000	+1.2
100	-19.1	2500	+1.3
125	-16.1	3150	+1.2
160	-13.4	4000	+1.0
200	-10.9	5000	+0.5
250	- 8.6	6300	-0.1
315	- 6.6	8000	-1.1
400	- 4.8	10000	-2.5

TABLE II

Relationships between sound pressure levels (SPL)
and sound power levels (PWL) (intensity)

(A) $PWL = SPL + 20 \log_{10} R + C \, dB$
     SPL measured at a distance R (metres) in the farfield of an
     omni-directional source on a flat reflecting surface

(B) $PWL = SPL + 10 \log_{10} A$
     SPL measured for a plane wave propagating over an area A, (m^2)

(C) $SPL = PWL + 10 \log_{10} \left(\frac{Q_\theta}{4\pi R^2} + \frac{4}{a} \right)$
     SPL of a reverberant enclosure with PWL of an enclosed source
     Q_θ is a directivity factor of the source
     a is the total absorption of the enclosure.

TABLE III

1.2   Characteristics of point aerodynamic sources

A general analysis of aerodynamic sources of sound is set out in the
Appendix A attached.   Firstly, suppose such a source is provided by an
unsteady discharge of fluid at an orifice with a velocity $v \propto V\exp(j\omega t)$.
Suppose further that the source has a characteristic dimension L, so that
typical frequencies will scale with velocity as V/L.   The fluctuating
volume flow giving rise to such a source will be proportional to $L^2 V$, so
that, using Equation A.11, we can show that p^2 varies with flow velocity
as $(\omega V L^2)^2$.   Thus the total acoustic power of the source will vary with

flow velocity in accordance with

$$W \propto 1.L^2V^4./4\pi\rho_o c_o \qquad\qquad\qquad 1.2(1)$$

The total energy flux or power associated with a steady flow of velocity V across an area L^2 is ρL^2V^3, so that this type of source has an acoustic efficiency that varies with velocity and is proportional to the flow Mach number $M = V/c_o$.

Unsteady relative motion between a body and a fluid (e.g. propeller or fan blade) gives rise to unsteady lift and drag forces. At the high Reynolds numbers, that normally exist in practical flow situations, aero-dynamic forces F are generally proportional to ρV^2L per unit breadth across the flow. Such fluctuating forces act acoustically in a way that is approximated by acoustic dipoles. Using Equation A.16 to evaluate p^2, one would expect the acoustic power per unit breadth to vary with flow velocity according to

$$W_b \propto \omega^2F^2L/\rho_o c_o^3 \quad \text{per unit breadth b,}$$

which gives

$$W \propto \rho V^6L^2/c_o^3 . \qquad\qquad\qquad 1.2(2)$$

For this case the acoustic efficiency will be proportional to V^3/c_ρ^3 or to M^3. This indicates that with subsonic flow, the relative efficiency of conversion from mechanical or flow energy to acoustic energy is much reduced relative to monopoles. Note also that the approximations of the theory need substantial modification at high flow Mach numbers and, in particular, as $M \to 1$.

Mixing noise produced by jets and wakes can be related acoustically to radiation by acoustic quadrupoles (See A1.0). With similar geometry, the magnitude of the turbulent velocity components v_i scales with the value of a characteristic flow velocity V. The associated frequency of the fluctuations again scales with the ratio V/L. Taking the corre-lated volume as L^3 and neglecting source convective effects, the acoustic power for such volumes of turbulent flow will, according to Equation A.20, be given by

$$W_v \propto \omega^4(\rho^2V^4)L^6/\rho_c c_o^5 \qquad\qquad\qquad 1.2(3)$$

so that

$$W_v \propto \rho V^8 L^2 / c_o^5$$

This shows that the acoustic efficiency is proportional to M^5 for mixing noise, for low or moderate flow Mach numbers at least.   At higher flow speeds due account must be taken of source convection and much more care taken in evaluating the integrals in A20.

Some acconnt can also be taken of the likely spectral distribution of the sound radiated from aerodynamic sources.   In this context, care is necessary to distinguish between discrete or broad band sound radiation. Many practical examples of unsteady or turbulent flow exist, however, where characteristic frequencies observed in the flow correspond to fixed multiples of the ratio V/L.   Thus for a given geometry, spectral chara- cteristics are most usefully described in terms of the Strouhal number S_n, where

$$S_n = \frac{fL}{V} , \qquad\qquad\qquad 1.2(4)$$

and f is the observed frequency.   Eddy shedding for a slender cylinder, for example, occurs at a Strouhal number of $0.2, |f = 0.2^V/d|$, for flow Reynolds numbers, based on diameter d, in excess of 100.

1.3   Aerodynamic or flow associated sources in the industrial
           environment

Many industrial and manufacturing processes employ compressed air (or other gases) for a wide range of factory operations.   Jets of com- pressed air are used for cooling, for moving components and for cleaning or swarf blowing.   Compressed air is also used for operating tools, presses or clamps and jacks.   The exhaust from such equipment is also a signifi- cant, if intermittent noise source.   Other sources comprise compressed air leaks or valve noise.   In all cases compressed air is normally supplied at around 5 atmospheres or more, so that most nozzles or other discharges are operating above the critical pressure ratio.

Other sources, which generally contribute more to factory background levels, rather than representing a direct hazard to health, are provided by fans, blowers, air extraction and ventilation equipment.   On the other hand, fan noise from ventilation or extract systems can often be a major contributor to noise created by industrial processes.   Fan noise represents a combination of thickness noise (or monopoles) at blade passing frequencies. dipole noise due to fluctuating forces on the blades and mixing or vortex noise from the blade wakes.

Statutory noise limits are specified in many ways. The simplest description of a noise source is its total effective intensity at some specified position, which is often measured with a dBA weighting network. For noise climate specification purposes, one may require a more detailed knowledge of the spectral distribution of the noise energy, in a contiguous set of either octave or third octave bands. Some standards require measurements of both linear (unweighted) or dB(C) overall sound pressure level, as well as the octave band A weighted levels. In some codes, limits are specified either in terms of overall permitted SPL in dB(A); in others they may be specified as a value of the N.R. noise rating index, which is calculated from octave band measurements of sound energy in each band from 31.5 Hz upwards.

The appropriate procedure to follow for predicting or estimating the source strength and the resultant noise climate, will depend to some extent on the code concerned. To meet most requirements what is required is, firstly, a set of guidelines or a procedure for estimating the total acoustic power of a given source and secondly, guidelines for making realistic estimates of the spectral distribution of that energy. The dominant aerodynamic sources in industry are generally associated with jet flows or high pressure blow off, or leaks. However, one can quickly demonstrate that the effective source strength of jets is magnified appreciably, when jets impinge onto fixed surfaces, giving rise to strong vortex shedding or vortex noise. This increases the acoustic efficiency of the source to approach that of a dipole or even that of a monopole.

Previous discussions have concentrated on the radiated sound field from a source region. However aerodynamic sound sources have a local fluctuating pressure field (sometimes termed pseudo-sound) which is associated with fluctuations in the flow. This is represented close to the source by the pressure term in the energy equation and is of the same order of magnitude as ρVv, where v is the fluctuating and V the mean component of the flow velocity. However, this pressure field decays rapidly with distance and only represents a hazard to an industrial operator, when such sources are relatively close to his ear. The practical extent of this near sound field is of the order of half a wavelength from the source. This fact should be taken into account during noise measurements. Nearfield pressure fluctuations or pseudo-sound can act as a generator of farfield acoustic radiation if it can excite a resonator. For example in the manner that sound is produced by blowing across the neck of an empty bottle.

1.4   Valve and blow-off or discharge noise

Though jet noise has been widely recognised as a major aerodynamic noise source in connection with aircraft, careful examination of the source characteristics of practical industrial jet flows indicate that the source strength is usually at least one order of magnitude above that expected from the jet mixing noise alone. It quickly becomes obvious that a control

valve or some other upstream mechanism is the major source.   A detailed
study valve noise (1.1) indicates that it is a complex problem involving
valve design and geometry, proportional valve opening and the pressure
drop across the valve.

For choked valves, Small (1.1) suggested that the acoustic efficiency
of fully choked valves of widely different geometry, running wide open,
could be described by $\eta = 10^{-4}kM^5$ with $0.65<k<1.35$.   This implies that
the source is quadrupole in nature if one records the discharge noise of
such valves.   The flow power and Mach number was evaluated at the fully
expanded flow velocity V.   This result may seem somewhat surprising, since
other measurements suggest sound power varies as V^6 not V^8, particularly
at lower flow rates.   This change is probably due to flow acoustic
interactions, while a sound pressure proportional to V^6 seems in better
agreement with most measurements with subsonic flow velocities through
the valve ports.

The spectrum of valve noise is generally broad band, though for some
flow conditions the valve may act as an acoustic resonator and produce
pure tones with harmonics.   The shape of a typical valve noise spectrum
is shown in the Figure 2.

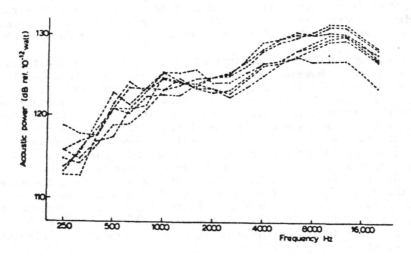

FIG. 2   1/3 Octave Band Spectra of Various Valves Passing Air From
        P(a) = 5 Bar to P(a) = 1 Bar at Q(st) = 2,700 m 3/h (Approx)

In this case the spectrum has two peaks, the higher frequency peak
relates to the valve port dimensions and the lower frequency to the pipe
dimensions.   Both correspond roughly to a Strouhal number of around 1/3.
The results correspond to seven quite different types of valve, including
a gate valve, of the same nominal size.

The levels and spectral distribution of blow-off (or venting) noise and discharge noise are very similar to those provided by valve noise. Resonance effects in discharge pipes etc can amplify some components of the spectrum to give such sources a tonal character.

References

1.1   D.J. Small. 1972. The noise of gas regulator valves.   The Institution of Gas Engineers, London, communication No.883.

1.2   H.D. Baumann. 1970. On the prediction of aerodynamically created sound of control valves. ASME WA/FE-28.   see also

American Gas Association publications e.g. LOO 230, LOO 280.

1.3   G. Reethof. 1978. Turbulence generated noise in pipe flows. Annual Review of Fluid Mechanics 10, 333-367.

1.4.   Fans and blowers (rotating machinery) noise

Propellers or fans, such as used on airplanes or helicopters, or in cooling, ventilating or exhaust systems can be strong sources of aerodynamic noise.   A fan or blower has two or more blades of finite thickness which carry with them a static pressure field as they rotate.   A microphone mounted near the plane of the fan will register a pressure peak for the passage of each blade.   The frequency of the peaks is the rotational speed times the number of blades and is called the blade passing frequency.

Such a disturbance of the air by the passage of a blade generates sound by volume displacement of the medium, which is equivalent to mono-pole radiation.   This provides a significant source of sound only when the blade velocities are high so that components of the induced pressure field have supersonic velocities (see A1.1).   Thus displacement flow noise is generated chiefly at the blade tips.   As the tip speed approaches the speed of sound more and more harmonics of the blade passing frequency will be generated.

The blades are producing a thrust on the air passing through the fan disc, while they also experience drag forces.   Any non-uniformity of the flow (flow inlet or exit distortion) will result in unsteady forces so that these also become dipole sources of sound.   Tones are produced when the flow irregularities result from the wakes of struts, guide vanes or splitters upstream.   Broad band noise will result from more disordered fluctuations in the inlet flow. Since the blades are moving, the steady aerodynamic forces on the fan or propeller blades also produce noise represented by a source distribution of dipole order in the propeller disc.

A further source of noise is produced by vortex shedding at the blade trailing edges and tips.   Noise results when the flow around the blade

is modified or modulated by the unsteady wake flow.   Even more dramatic
is the result of recirculation when shed vortices interact with succeeding
blades.   Blade slap, a prominent feature of helicopter noise, is
produced when one blade passes through the strong tip vortex of a
neighbouring blade.

Interaction between moving blade rows also produces noise and are a
significant factor in the noise produced by multistage fans and compressors.   The periodic forces depend on the flow distortion imposed by one
row or another.   Centrifugal blowers also generate noise which is
related to interactions between the flow in neighbouring passages in the
runner and the flow in the volute.

Large low speed fan noise can be predicted roughly for the sound
intensity in the blade passing frequency octave band by

$$PWL = 125 + 20\log_{10}HP - 10\log_{10}Q(ft/min)$$

This relates to a fan running at or near design conditions.   With centrifugal fans the overall sound level is also dominated by the blade passing
tone.   Results vary to some extent but the variation of sound power is
well described by

$$W \propto u^{5.6}D^{2.4}$$

where u is the tip speed and D the impeller diameter.   Piestrup and
Wesler suggest another correlation for large blowers, or

$$PWL = L05 + 17.7\log_{10}(HP/N) + 15\log_{10}(N/6)$$

where HP is horse power for stage and N is number of blades.

Electric motor cooling fans also produce noise which depends on the
amount of cooling air required and the size and shape of the air passages.
The noise level thus depends on details of the motor design.   There are
design standards for motors which include noise, i.e. BEAMA 225, VDE 0530
(German) and NEMA MG1 - 12.49, 1970.

References

1.5.   C.F. Piestrup and J.E. Wesler, 1963.   Noise of ventilating fans.
       J.A.S.A. Vol.25 Extensive bibliography (190 references in:-
       C.L. Morfey 1973.   Rotating blades and aerodynamic sound.   J.Sound

Vibration 28, pp 587 - 617.

1.6   Sources associated with cyclic flow

Another group of noise sources are associated with equipment producing a cyclic flow.   With gas flows these include positive displacement blowers, reciprocating compressors, internal combustion engines,
etc.   Most of these devices are so noisey that some form of intake or
exhaust silencing is normally required.   With liquid flows, this includes
a wide range of hydraulic pumps, (e.g. gear pumps) and machinery, with a
separate group of problems where cavitation provides a significant noise
source.

Rootes blowers are particularly noisy and produce strong tones at
the lobe passing frequency and its harmonics.   A useful prediction for
two-lobe Rootes' blowers can be made from

in pipe SPL - 50 $\log_{10} u_t$ + 21 dB.,

where u_t is the tip speed in ft/sec.

The low frequency source characteristics of reciprocation compressors
comprise a set of harmonics of the pulse repetition rate.   This is equal
to the rotational speed N for single acting and twice the rotational speed
for double acting compressors.   Since inlet and exhaust ducting is provided
with multicylinder machines, the fundamental of the harmonic series
remains unchanged, though the relative strengths of higher harmonics will
alter as more cylinders are added.

Similarly the low frequency intake and exhaust noise of internal
combustion engines consists of a set of harmonic tones with a fundamental
frequency equal to the firing rate of one cylinder (i.e. N/2 for four
stroke and N for two stroke).   The relative magnitude of the harmonics
varies with the number of cylinders, manifold and exhaust pipe geometry,
valve cam profile and to some extent on valve overlap.

High frequency flow noise is invariably generated in all cyclic flow
systems.   Except for cavitation in fluids, it is not of major significance
unless either the pipe flow velocity exceeds 0.1 Mach number. or the
unsteady flow excites an acoustic resonance in the inlet or exhaust duct.

A limiting case of cyclic flow is when the fluctuation is effectively
confined to a single pulse.   Blast waves or explosion waves are generated
on the sudden opening of pressure relief valves or explosion door.   The
discharge normally takes place through a vent or duct.   The source can
therefore be complex since the initial explosion or blast waves is followed
by a decaying train of acoustic waves produced by a transient standing

wave in the duct.   The pressure rise in the blast wave front is sudden,
and represents energy at the high frequency end of the spectrum.   This
decays at a significantly higher rate than the low frequency reverberant
oscillations excited in the duct.   Thus, even at relatively modest
distances (100 m say), the low frequency reverberant component provides
most of the transmitted sound energy.

References

1.5   T. Priede, 1967. Origins of rotary pressure displacement blower
       noise. I. Mech. E. London conference "Noise from power plant
       equipment."

1.6   R.J. Alfredson, P.O.A.L. Davies, 1970. The radiation of sound from
       an engine exhaust. J. Sound Vibration 13, pp 389 - 408.

1.7   P.O.A.L. Davies, T.B. Guy, 1970. Shock wave propagation in ducts.
       Proc. I. Mech. E., London, 184, pp 120 - 124.

2.0   Aircraft noise mechanisms

      The development of high speed jet propelled aircraft brought with it
a host of aerodynamic noise problems.   Propeller noise had received some
attention prior to 1940, but experimental and theoretical studies of aerodynamic noise obtained a substantial boost in the 1950's, with the development of the civil jet aircraft   The problem has two aspects, first
the social impact of aircraft noise with its political implication and
secondly the technological implication concerned largely with acoustic
fatigue of aircraft structures.

      The major source of aircraft noise is provided by the propulsion
system   This includes the jet efflux, tones from the fan or compressor
and combustion noise.   Other sources which may be of equal significance
in appropriate circumstances are boundary layer noise, noise or pressure
fluctuations from vortex shedding and resonant excitation of cavities.
Major effort has been concentrated on the various sources associated with
the propulsion system, with a gradual accumulation of effective noise
control technology as understanding of the basic mechanisms improved.
Much of the understanding and knowledge accumulated in propulsion noise
studies has direct application to aerodynamic noise problems in other
fields.   The same is true with regard to other mechanisms that are
associated with aircraft noise.

      A general review of aerodynamic noise sources and an outline of the
mechanisms involved has been given in Section 1 of these notes.   Further
details of the mechanisms are set out here.   It is worth mentioning that,
though described in the general context of aircraft noise, the field of
application is much wider than this.

2.1   Propulsion noise

Propulsion noise embraces several component mechanisms which will be considered separately.   The most fundamental, and in practice the most difficult to eliminate, is that due to turbulent mixing of the hot jet exhaust with the ambient fluid downstream of the nozzle exit plane. Incorrectly expanded jet exhaust flows give rise to shock waves which then give rise to shock associated noise.   This appears as discrete tones, referred to as screech, together with broad band radiation, or broad band shock-associated noise.   The development of high by-pass ratio engines has highlighted other noise sources.   These are referred to as excess noise, tailpipe noise or core noise, though the precise origins of such noise is not yet established.   Uncertainty exists as to whether it arises within the engine, due to unsteady combustion, flow over support structures, etc., and subsequently passes out with the jet efflux as acoustic energy, or whether it arises as the result of interactions between vorticity and the nozzle termination.   Current evidence suggests the former mechanism is dominant in practice.

Figure 3 illustrates the spectral distribution of the noise from an early by-pass engine, indicating the relative importance of the various components.   The diagram also indicates the relative changes brought about by developments intended to reduce the noise.

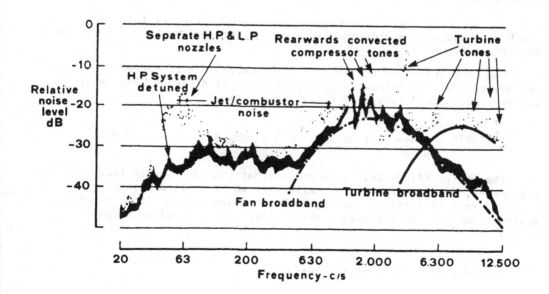

FIG. 3   Compressor and turbine component sources

Current propulsion noise prediction methods are largely empirical, mainly due to the fact that excess or tailpipe noise provides a dominant source in contemporary production high by-pass engines. We shall concentrate here, however, on sources whose mechanisms are sufficiently well understood to provide valid predictions of the radiated noise.

2.1.1 Jet mixing noise

We saw earlier, in 1.2 Equation 1.2(3) that the far field noise intensity of a cold jet will vary dimensionally as

$$I(r,\theta) \propto \frac{\rho_s^2}{\rho_o} \frac{U_j^8}{c_o^5} \frac{D^2}{r^2}, \qquad\qquad 2.1$$

where U_j is the jet efflux velocity, D its diameter, ρ_s the density of jet flow and r the distance to the observation point. This applies to 'point' quadrupole radiation, that is, there is an assumption that the dimensions of the source region are small compared with the acoustic wavelength of the sound. With sources in motion (See A1.4) the equation above is modified by the further factor $(1-M\cos\theta)^{-5}$ where M is normally of order $0.7U_j/c_o$. This factor leaves unaltered the noise radiated normal to the jet flow direction, but augments or reduces the noise radiated in other directions.

The spectral characteristics of jet mixing noise for full radiation are broad band with a wide peak at around a Strouhal number of 1, as can be seen in Figure 4.

When the intensity is measured in frequency bands, however, Lush 2.1 has argued that the intensity in proportional frequency bands, centre frequency f, should vary at a given angle λ as

$$I(f,r,\theta) \propto \frac{\rho_s^2}{\rho_o} \frac{U_j^8 D^2}{c_o^5 2} \left|1-M\cos\theta\right|^{-5} F(\frac{fD}{U}(1-M\cos\theta)). \qquad 2.2$$

The factor $F(1-M\cos\theta)$ is to ensure that the same source frequency is considered, irrespective of the angle from which it is observed. Also with aircraft propulsion systems, jet velocities are such that $M\cos\theta$ can approach unity. To avoid this singularity a more complete Doppler factor

$$\{(1-M\cos\theta)^2 + (\alpha^2\sin^2\theta + \beta^2\cos^2\theta)M^2\}^{\frac{1}{2}}$$

should be employed. (2.2).

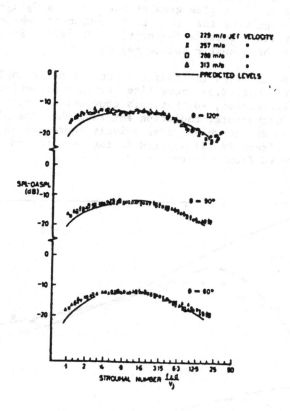

FIG. 4. Third-octave relative spectra measured
under static conditions

Comparison with experiment showed, however, that the theory of freely
convecting quadrupole sources, represented by equation (2) above, fails
to predict the observed directivity. The discrepancies are small at the
lowest Strouhal number {0.1}, but increase with Strouhal number. The
discrepancy has been explained in terms of the known property of a wind
shear in refracting sound, during its propagation through the atmosphere.
The mixing region of the jet is highly sheared flow, which has the property
of refracting the sound away from the jet axis.

Taking such refraction into account, using ray acoustics, one finds
that there is a critical minimum angle {to the flow direction} within
which no sound is radiated to the far field. This angle provides the

boundary to a 'cone of silence.'   It is not difficult to show that
propagation only occurs at angles greater than arc $\cos(c_s/c_0+M)^{-1}$.   This
is equivalent to recognising that only those components whose phase speed
exceeds c_s+v_s can radiate outside the critical angle.   (See, e.g. A1.1).
Here c_s+v_s refer to the acoustic source region.

Refraction also modifies the radiation at all other angles.   Careful
study of the problem Ref.(2.3) shows firstly, that directivity is
modified by such diffraction, so that this convective factor is modified
to $(1-M\cos\theta)^{-3}$.   Furthermore, radiation entering the cone of silence will
be attenudated by an amount which progressively increases as the angle θ
is reduced.   This effect is illustrated in the results and comparisons
in Figure 5 reproduced from reference (2.3).

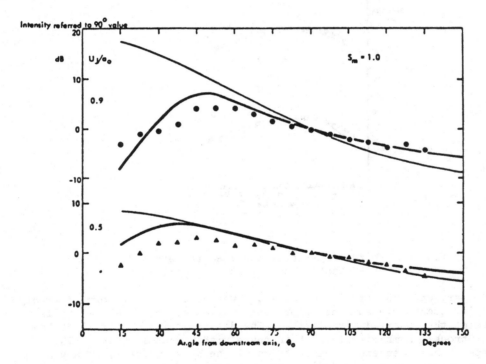

FIG.5   Directivity of Isothermal Jet Noise, Comparison of the
        Critical Models with 1/3 Octave Data, for Sm = 1.0 points
        measurements, calculations from reference (2.3)

Due account must also be taken of the fact that the jet flow is
hotter than the surroundings.   Heating the jet will reduce the flow
density ρ_s (See equation 2.2) leading to a reduction of noise radiation
with rise of temperature.   The problem is, however, more complicated

than this (2.4).   The analysis presented first by Lighthill neglects
several effects, in particular the fact that entropy must be conserved in
a frame of reference following a fluid particle.   Including the fluctuating entropy in the Lighthill stress tensor T_{ij} (see A1.0), one finds that
the sound radiation is expressed by [2.5].

$$I \propto \frac{\rho_s^2 \, U_J^8 \, D^2}{\rho_o c_o^5 \, r^2} + K \frac{\rho_s^2 \, U_J^6 \, D^2}{\rho_o c_o^3 \, r^2}(\log_e \frac{T_J}{T_o})^2 \qquad\qquad 2.3$$

where T_J is the jet flow temperature.   An alternative expression with a
U_J^4/c_o variation for the second term also fits the measurements, as shown
by the results in Figure 6 taken from (2.6).

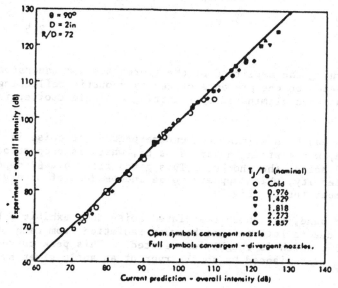

FIG.6   Comparison of Experiment with Current Prediction
         Model: Overall Intensity at $\theta = 90^\circ$.

2.1.2 Shock associated noise

When the pressure ratio across a convergent nozzle exceeds the
critical value (1.8 for air), a series of shock cells form in the exhaust
flow.   Further increase in pressure ratio extends the length and spacing
of successive cells, the spacing being given approximately by

$$L = 1.8\beta \, D \qquad\qquad 2.4$$

where $\beta = (M_J^2-1)^{\frac{1}{2}}$, and M_J is the fully expanded jet Mach number. Two types of noise radiation are associated with this shock cell structure, discrete tone radiation or screech (2.7), originating from a feedback mechanism and broad band radiation (2.8), arising from shock interaction with eddies carried by the flow.

The screech mechanism involves disturbances leaving the nozzle, interacting with the shock to produce acoustic radiation, which returns via the ambient air to initiate a new disturbance. The downstream travel time is L/V_c, where V_c is the convection velocity of the disturbance while the return time is L/C_0. This gives a fundamental frequency of the pure tones f_s as

$$f_s = \frac{V_c}{L(1+M_c)} .$$
                                                                        2.5

The factors governing the amplitude of the source are not understood, but it apparently depends on the presence of nearby acoustic reflecting surfaces. It has been eliminated by inserting a single spoiler in the flow.

When screech has been eliminated, underexpanded jet noise is dominated by another source, whose strength varies as β^2, which is proportional to the pressure jump across the shocks. This gives rise to a β^4 dependence for the sound intensity, which appears to be independent of observation angle and jet temperature. (Fig.7).

Though broad band, the shock associated noise does exhibit a broad peak that arises due to interference between radiation from each of the row of shock cells, which is roughly correlated. This peak corresponds to a frequency f_p, experienced by an observer at an angle given by

$$f_p = \frac{V_c}{L(1-M_c \cos\theta)}$$
                                                                        2.6

This differs distinctly from the screech frequency defined in 2.5. A comparison with this prediction and the spectrum shape when θ is 90° can be seen in Figure 8.

2.1.3 Fan or compressor noise

The fan or compressor also makes a substantial contribution to the propulsion system noise. The source mechanisms involved with fan noise have been discussed earlier in Section 1.5. In the case of aircraft propulsion systems the straight jet and the by-pass, or turbo-fan engine, represent two distinct problems. During the course of development, initially, the jet with combustion or excess noise was the dominant source.

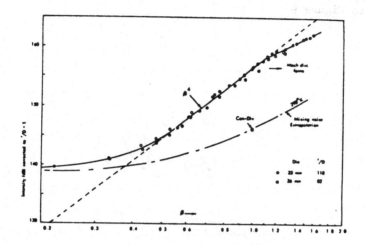

FIG.7 Variation of Overall Intensity at 90° to
Jet with $\sqrt{\pi^2-1}$

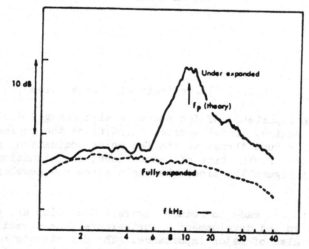

FIG. 8 Comparison of Supersonic Jet Noise Spectra for
a Fully Expanded and Underexpanded Flow($\theta=90^\circ$, $\beta = 1.0$)

With progression to higher by-pass ratios, fan noise has become troublesome, both in the forward and rearward areas. The relative importance
of fan or compressor noise as a major source of propulsion noise is
illustrated in Figure 9.

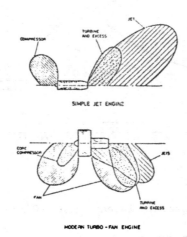

FIG. 9 Ingredients of Engine Noise

The relative importance of fan noise is also changed during normal
operating of the engine. Under approach conditions the engine is
throttled back to reduce thrust so the fan noise domination, particularly
for the forward arc. During take off conditions, the situation changes,
as can be seen in Figure 10, since the engines are now developing full
thrust.

Great efforts are made to reduce aircraft fan noise and compressor
noise during design and development. Furthermore, the directivity of
the radiation is also of major importance. The directivity of open
rotors can be predicted more or less satisfactorily, but such predictions
with shrouded (fans) or ducted (compressors) rotors are much more difficult
The duct geometry and duct wall lining characteristics have a pronounced
effect on sound propagation, while the flow field and geometry at the duct

entrance have a major effect on directivity at different frequencies. With a ducted rotor, the rotating sources excite higher order acoustic duct modes, only some of which propagate along the duct, (i.e. those with appropriate phase velocities in the direction of propagation).

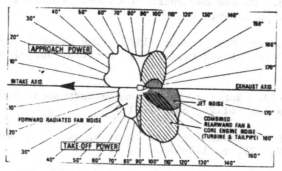

FIG. 10   Fan Engine Noise Sources

For a classical, one dimensional hard-walled rectangular duct, the condition for modes to propagate unattenuated can be shown to be

$$m \, B \, M_e > q$$

where q is the order of the source mode, m the sound pressure harmonic, B the number of blades and M_e the effective blade Mach number.   Further information can be found in a review paper by Morfey (2.9) and papers by Wright (2.10) and Mugridge (2.11).   The characteristics of propagation in flow ducts is treated later, but see (2.12).

References

2.1   P.A. Lush 1971. Measurement of subsonic jet noise and comparison with theory. J.Fluid Mech. Vol.46, pp 477-500.

2.2   J.E. Ffows Williams. 1963. The noise from turbulence convected at high speeds. Phil.Trans. Royal Society. A255, pp 469-503.

2.3   B.J. Tester and C.L. Morfey. 1976. Development in jet noise modelling and theoretical predictions and comparisons with measured data. J. Sound and Vibration, Vol.46, pp 79-104.

2.4   M.J. Fisher, P.A. Lush and M. Harper-Bourne. 1973. Jet noise. Journal Sound and Vibration Vol.28, pp 563-585.

2.5   C.L. Morfey 1973. Amplification of aerodynamic noise by convected flow inhomogeneities. Journal Sound and Vibration Vol.31, pp 391-397.

2.6 H.K. Tann, P.D. Dean and M.J. Fisher, 1975. The influence of
     temperature on shock free supersonic jet noise. Journal Sound and
     Vibration Vol.39, pp 429-460.

2.7 A. Powell. 1953. On the mechanism of choked jet noise. Proc.Phys.
     Soc. B. Vol.66, pp 1029-1056.

2.8 M. Harper-Bourne and M.J. Fisher, 1973. The noise from shock waves
     in supersonic jets. AGARD Conference pre-print No.131 on noise
     mechanisms.

2.9 C.L. Morfey, 1973. Rotating blades and aerodynamic sound. Journal
     Sound and Vibration Vol.28, pp 587-617.

2.10 S.E. Wright, 1972. Waveguides and rotating sources. Journal Sound
     and Vibration Vol.25, pp 163-178.

2.11 B.D. Mugridge, 1969. The measurement of spinning acoustic modes
     generated in an axial flow fan. Journal Sound and Vibration Vol.10,
     pp 227-246.

2.12 P.E. Doak, 1973. Fundamentals of aerodynamic sound and theory and
     flow duct acoustics. Journal Sound and Vibration Vol.28, pp 527-561.

2.2 Airframe noise

Noise is generated by the interaction between the flow and the surfaces
of an aircraft.   Here it is normal to distinguish between the clean, or
normal flight cruise condition and the 'dirty' or normal landing configuration.   The major mechanisms involved include the boundary layer noise,
lift and drag noise, eddy shedding and wake vorticity noise, buffet and
eddy shedding from struts and wheels, noise from cavities (i.e. wheel
wells) and cut outs.   Other sources arise from the impingement of jets
on flight surfaces or the passage of strong vortices past trailing edges.
A typical set of such sources in the landing configuration is illustrated
in Figure 11.

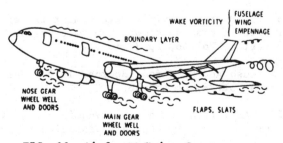

FIG. 11 Airframe Noise Sources

2.2.1 Boundary layer and flight noise

Boundary layer noise is generated by the turbulent pressure fluctuations in the boundary layers existing on all aircraft surfaces in flight.   It is not a significant contributor to the far field noise except (a) where the surface responds to the fluctuating pressure field to amplify the sound or (b) at leading or trailing edges where the boundary layer turbulent eddies can radiate strongly as they pass the discontinuity. A major practical problem exists when lightly loaded structures are excited by the boundary layer pressure fluctuations, since this can produce fatigue damage of the structure.   Transmission of the boundary layer pressure fluctuations to the cabin interior represents another noise problem.

The wall pressure field is produced by contributions from pressure sources in the boundary layer with a wide range of convection velocities. Its magnitude lies between 0.5 and 1% of the flow dynamic head.   On the basis of Strouhal simularity, two families of wave-number components can be identified.   High wave number components are associated with the sheared flow at the surface and contribute three-quarters of the mean square pressure field.   They remain strongly correlated for four or five wavelengths, in the streamwise direction and for about one-third of this in the cross-stream direction.   The low wave number components are associated with large scale eddy motions in the outer edge of the layer.   The intensity of excitation of the structure depends on the scale over which the pressure forces remain coherent, relative to characteristic structural wavelengths, as well as on their phase velocity, relative to the flexural wave speeds.

Measurements of the far field radiated noise from aircraft in flight involves several practical difficulties, but the results indicate that the major source is dipole in character and arises from the integrated effect of the fluctuating forces acting over the aircraft surfaces.   Such unsteady aerodynamic forces arise from inflow turbulence, from the boundary layer, or from the turbulent wake interacting with the surfaces.   Most experimental evidence suggests that the dominant source mechanism is the interaction of the turbulent wake flow with the airfoil to produce unsteady lift and drag forces giving rise to low frequency noise.   The same interactions generate relatively small scale pressure fluctuations at the trailing edge (vortex shedding) producing high frequency noise.   The directivity pattern for the former corresponds to a point dipole and for the latter to a baffled dipole as shown in Figure 12.

Since observations suggest that the noise is dipole in character and should thus be proportional to $V^6 S/r^2 A^4$,

i.e. $\text{OSPL} = 10 \log_{10} \left| \dfrac{V^6 S}{r^2 A^4} \right| + K_1,$ 2.7

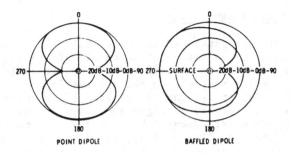

FIG. 12.   Directivity of airframe noise

where S is the surface area, r the distance and A the aspect ratio.
Regression analysis of 53 measurements, however, gives

$$OSPL = 10 \log_{10} \left| \frac{V^{4 \cdot 93} S^{0 \cdot 72}}{r^{1 \cdot 62} A^{2 \cdot 06}} \right| + K_2 \ . \qquad\qquad 2.8$$

The differences suggest that the sources are probably not compact.
The noise spectrum is broad band but peaks at a Strouhal number, based
on wing thickness, of around 1.3.

2.2.2 Landing configuration noise

In the approach or landing configuration, the character of the airframe noise changes.   Flight velocities are lower, but the aircraft is
operating at high lift coefficients so that there are relatively more
regions of separated or unsteady flow.   Thus more mechanical energy is
being dissipated by drag.   Thus the broad band dipole noise would
represent a source strength given by

$$OSPL = 10 \log_{10} \left| \frac{C_D^2 V^6 S}{r^2} \right| + K_3 \qquad\qquad 2.9$$

where C_D is the drag coefficient.   This should be compared with the two
previous equations where the similarity is evident.   For example the
induced drag produces the dominant noise with a 'clean' lifing wing.
With C_{Di} roughly proportional to A^{-1}, so that 2.8 and 2.9 are consistent
with each other.

The wheel wells and other cavities provide resonant chambers which
can be excited by the unsteady flow past them.   Such resonant fluctuations
may be amplified by a feed back mechanism to produce a significant monopole

source of tonal character.   The basic mechanisms and characteristics of
such sources will be discussed later in Chapter 3.   The difference in
airframe noise source strength between clean (flight) and dirty (approach)
configurations amounts in total to around 5 - 6 dB.

References

H.G. Morgan, Airframe noise - the next aircraft noise barrier.

J.C. Hardin, 1974.   AIAA Paper No.74 - 949.

J.S. Gibson, 1972, 1973.   Papers in Internoise proceedings.

P. Fethney, 1975.   An experimental study of airframe self noise.
R.A.E. Technical Memo. Aero.1623.

M.K. Bull, 1967.   Wallpressure fluctuations associated with subsonic
turbulent boundary layer flow.   Journal Fluid Mechanics, Vol.28,
pp 719 - 754.

3.0   Sound generation and propagation in flow ducts

Flow ducts provide another important and widespread group of aerodynamic noise problems.   One group of sources are provided by steady or
cyclic discharge through control valves or ports.   Another by unsteady
pressures and flows from rotating machinery.   Some of the basic mechanisms
concerned with such sources have been discussed already.   Yet another
class of sources are generated by edge tone mechanisms, these involve
unsteady or periodic eddy shedding, the phenomenon being controlled or
modified by feed-back and amplified by resonance.

The study of flow duct noise problems includes (1) source identification and characterisation; (2) acoustic energy propagation along flow
ducts, including the effect of discontinuities and various forms of non-
linear behaviour; (3) sound radiation from flow intakes or discharges and
(4) the radiation of sound from the duct walls.   The study of sources
includes the special group of high intensity acoustic sources that are
provided by sirens of various types.   These represent particularly
efficient and effective converters of mechanical to acoustic energy.

Experimental and theoretical studies of flow duct acoustics are best
approached on the basis of transmission line equations.   In their simplest
form, they describe the propagation of plane waves in a tube with
acoustically hard walls, where the diameter is small compared with the
wavelength.   More complex relations are needed to describe more complex
wave motions, involving higher order modes of wave motion.   Any quantitative description of energy propagation also involves a careful specification of the boundary conditions.   Non-linear behaviour exists when
flow is present, for certain classes of discontinuity or boundary conditions,

or when the pressure amplitude is high. (of order 0.1 bar or more).
For simplicity, we shall consider only plane mode propagation specifically,
in so far as the analysis presented here is concerned.   Where appropriate,
however, situations requiring more detailed analysis will be identified
and an outline of the basic approach laid down.   A recent detailed
general theoretical analysis by Doak of flow duct acoustics can be found
in (3.1) and (3.2).

References

3.1 P.E. Doak, 1973. Excitation, transmission and radiation of sound
     from source distribution in hard walled ducts of finite length.
     (i) The effect of cross-section geometry and source distribution
     space-time patterns. Journal Sound and Vibration, Vol.31,pp 1-72.

3.2 P.E. Doak, (ii) The effect of duct length. Ibid 137 - 174.

3.1   Plane wave propagation in ducts, zero flow

(a) Ducts of constant section

Acoustic energy transmission in ducts is by a wave mechanism, which
we will assume remains substantially one dimensional {plane waves} and
steady.   The energy is provided by a source, which excites the wave
motion.   Any discontinuity in the duct (i.e. change of geometry, bend
or change of wall impedance) will modify the wave motion, some of the
energy being transmitted as a new wave, the remainder being reflected as
a reflected wave.

The normal situation is, therefore, to find two sets of waves at
any transverse plane along the duct;   the incident wave travelling out
from the source at a velocity c, relative to the moving gas, and the
reflected wave returning towards it.   With the broad band excitation
that aerodynamic sources produce, there will always exist a set of
standing waves in flow ducts, produced by the summation of particular
sets of incident and reflected waves.

The incident wave is conveniently described by

$$\hat{p}_i = \hat{p}_i e^{i(\omega t - kx)} e^{-\alpha x} , \qquad\qquad 3.1$$

$$\hat{v}_i = \frac{\hat{p}_i}{Z_s} e^{i(\omega t - kx)} e^{-\alpha x} , \qquad\qquad 3.2$$

where $\hat{p}_i$ and $\hat{v}_i$ are the (complex) pressure and velocity amplitude, ω the
radian frequency k the wave number ω/c and α a coefficient which

represents decay of wave energy as it propagates along the duct.
Similarly the reflected wave is described by

$$\hat{p}_r = \hat{p}_r e^{i(\omega t + kx)} e^{\alpha x} \quad , \qquad\qquad 3.3$$

$$v_r = -\frac{\hat{p}_r}{Z_s} e^{i(\omega t + kx)} e^{\alpha x} \quad , \qquad\qquad 3.4$$

An alternative description is to express the pressure etc. by $\hat{p} e^{i\omega t} e^{-\gamma x}$
for positive going waves, where $\gamma = \alpha + i\beta$. With hard walled ducts and
zero flow, then $\alpha \to o$, $\beta \to k$ while the duct impedance $Z_s = \rho c$, the characteristic
acoustic impedance. The total sound pressure is then given by

$$p = p_i + p_r \qquad\qquad 3.5$$

and the corresponding expression for the particle velocity is then

$$v = v_i + v_r \quad . \qquad\qquad 3.6$$

Neglecting, for the moment, attenuation $\alpha(\alpha \to o)$, given that the
pressure and velocity at $x = o$ is p_o and v_o, at any other point ℓ, with
$\ell = -x$ one finds that

$$\hat{p}_\ell e^{i\omega t} = \left| \hat{p}_o \cos k\ell + i Z_s \hat{v}_o \sin k\ell \right| e^{i\omega t} \quad , \qquad\qquad 3.7$$

and

$$\hat{v}_\ell e^{i\omega t} = \left| \frac{i\hat{p}_o}{Z_s} \sin k\ell + \hat{v}_o \cos k\ell \right| e^{i\omega t} \quad . \qquad\qquad 3.8$$

(b)   Ducts with prescribed boundary conditions

Particular cases of interest exist for prescribed values of the
boundary conditions at $x=o$.

Firstly, if at $x=o$, $v=o$ (a hard termination) then

$$\hat{p}_\ell = \hat{p}_o \cos k\ell; \quad \hat{v}_\ell = (\hat{p}_o / Z_s) i \sin k\ell \quad . \qquad\qquad 3.9$$

We see that this represents a standing wave with velocity leading
pressure by a phase angle $\pi/2$. {a quarter wavelength}.   With a tube
cross section A, the volume velocity at ℓ is Av_ℓ so the acoustic impedance
Z_{oc} is given by

$$Z_{oc} = \frac{Z_s}{iA} \cot(k\ell).$$ 3.10

The relations 3.9 and 3.10 are illustrated in Figure 3.1

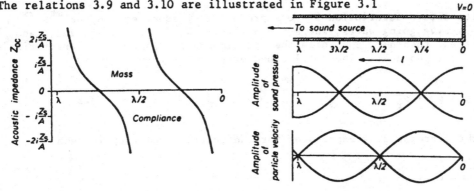

FIG.   3.1

We note in particular that the acoustic impedance is purely imaginary,
the magnitude of the input impedance being indicated in the figure.

        Alternatively, if the prescribed boundary condition at x=o are that
p=o, (an ideal open end) then,

$$\hat{p}_\ell = Z_s\hat{v}_o i \sin k\ell; \quad \hat{v}_\ell = \hat{v}_o \cos(k\ell) ,$$ 3.11

with

$$Z_{sc} = i(Z_s/A)\tan k\ell \quad .$$ 3.12

These results corresponding to α = o are illustrated in Figure 3.2

We note that, in this example, the acoustic impedance becomes infinite
when the tube length is an odd multiple of a quarter wavelength.   The
acoustic impedance is very much greater than the steady flow resistance,
so that very high pressures are required to generate an alternating flow.
This fact should be taken into account when interpreting measurements

made in flow ducts, e.g. of i.c. engine exhaust noise. Thus the measured sound is strongly modified by the effective impedance of the duct at each frequency.

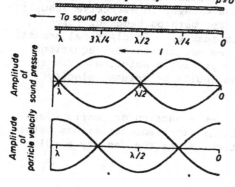 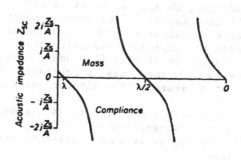

FIG. 3.2

A more realistic case is when the boundary conditions at x=o are specified as an impedance Z_w. In this case some of the incident wave energy will be reflected and some transmitted. In particular the ratio (usually complex) of the reflected to incident wave amplitude termed the reflection coefficient r is determined by

$$r = \frac{\hat{p}_r}{\hat{p}_i} = \frac{Z_w - Z_s}{Z_w + Z_s} = Re^{i\phi} \quad . \qquad\qquad 3.13$$

Making this substitution, 3.7 and 3.8 become

$$\hat{p}_\ell = \hat{p}_i \left| e^{-ik\ell} + Re^{i(k\ell+\phi)} \right| \quad , \qquad\qquad 3.7(a)$$

and

$$Z_s \hat{v}_\ell = \hat{p}_i \left| e^{-ik\ell} - Re^{i(k\ell+\phi)} \right| \quad , \qquad\qquad 3.8(a)$$

which have more general application.

Equation 3.13 is more representative of the proper boundary condition for the open end of a duct. It is well known that there is a frequency dependent phase change on reflection at the open end of a circular pipe (of an organ pipe) which increases its apparent length by 0.6 of the pipe radius. This exists because the pressure does not adjust immediately, since the fluctuating flow in and out of the pipe possesses inertia. This is considered again later in 3.3.

(c)   Discontinuous ducts

Most ducts in practice include discontinuities such as bends, junctions, changes in cross section area and changes in effective acoustic wall compliance.   Each discontinuity - except perhaps bends of large radius - represents a change of acoustic impedance.   In classical acoustic theory the impedance was calculated using the one dimensional equations of continuity of mass flow and of energy to provide a solution.   It is easy to show that this is equivalent to continuity of volume velocity and of pressure at the discontinuity.

A common discontinuity is a simple sudden expansion or contraction in duct cross-section area.   If p_1, v_1 and A_1 represent conditions before and p_2 etc., after such a discontinuity, then, since the density hardly changes,

$$\hat{p}_{1_i} + \hat{p}_{1_r} = \hat{p}_{2_i} + \hat{p}_{2_r} \ , \qquad\qquad 3.14$$

but see 3.17.   Also we have for continuity of volume velocity

$$(\hat{p}_{1_i} - \hat{p}_{1_r})A_1 = (\hat{p}_{2_i} - \hat{p}_{2_r})A_2 \ . \qquad\qquad 3.15$$

Comparison with experiment shows that this result is an approximation. The proper boundary conditions require zero velocity over the surface of discontinuity that provides the area change from A_1 to A_2, and this condition is not expressed by 3.15.   Continuity of volume velocity may however be closely approached a short distance either side of the discontinuity.   In general it can be shown that most discontinuities give rise to higher order modes and that these are required to satisfy the boundary conditions explicitly.   In many cases these modes are non-propagating and decay rapidly as one moves away from discontinuity.

The appropriate analysis required for such situations has recently been set out by Cummings, for some representative examples, being reported in references (3.3) and (3.4).   Inadequate specification or treatment of boundary conditions may not be serious for many applications, where the errors involved in applying such simplified relations as 3.14 and 3.15 may be insignificant, compared with those errors arising from other uncertainties or inadequate data.   At area changes, for example, close agreement with observation is obtained if an end correction is added to the narrower duct.   In most practical cases it is less than 0.4 of the radius so may be negligible.

References

3.3 A.J. Cummings. 1974. Sound transmission in curved duct bends.
     Journal Sound and Vibration, Vol.35, pp 451 - 477.

3.4 A.J. Cummings. 1975. Sound transmission in a folded annular duct.
     Journal Sound and Vibration, Vol.41, pp 375 - 379.

3.2   Sound propagation in ducts carrying flow

(a)   Wave propagation in constant area flow ducts

The presence of flow modifies the propagation of sound in ducts and alters the relations for acoustic energy propagation set out above. The sound is now propagating through a moving fluid, so that its effective phase velocity, in terms of displacement along the duct, is increased in the direction of flow and reduced in the opposite direction. See Ref. 3.5. Such effects of flow are best represented in terms of a modified wave number.

With a mean flow velocity V, assumed positive with respect to the co-ordinate x the effective phase velocity in the flow direction is $c_o + V$. Since the wavenumber k relative to the gas is ω/c_o the effective wavenumber $k^+ = \omega(c_o + V) = k/(1+M)$. Similarly for the wave travelling against the flow $k^- = k/(1-M)$.

For ducts with flow, we find therefore the k^+ replaces k in equations 3.1 and 3.2 while k^- replaces k in 3.3 and 3.4. The Mach number M is that of the mean duct flow. A similar agreement leads to $\alpha^+ = \alpha/(1+M)$ in the same pairs of equations.

(b)   Uniform flow ducts, with prescribed boundary conditions

Sound radiation from the open end or discharge of a flow duct is a common practical situation. Assuming for the moment that any attenuation of sound in the duct is negligibly small and that conditions at the open end are described by 3.13, then the amplitude of the sound pressure is given by

$$\hat{p}_x = \hat{p}_o(e^{-ik^+x} + Re^{i\phi}e^{ik^-x}) \ ,$$

$$= pe^{i(k^- - k^+)x/2}\left|e^{-i(k^+ + k^-)x/2} + Re^{i\phi}e^{i(k^+ + k^-)x/2}\right| \ , \quad 3.7(b)$$

which can be compared with 3.7 for ducts without flow. This shows that the distance between the nodes of the standing waves (see Fig. 3.2) is reduced by the factor $2k(k^+ + k^-)^{-1}$ or $1-M^2$, with flow present. Measurements of the pressure reflection factor for a baffled opening (i.e. in an acoustically hard infinite wall), taken from (3.6), are shown in Fig. 3.3 (note $r_p = R$).

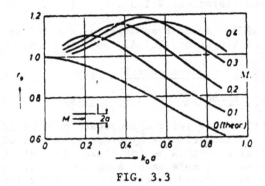

FIG. 3.3

An alternative example in Figure 3.4, taken from reference (3.7), shows measurements of R and ϕ (plotted as θ) for an unbaffled or unflanged pipe. The phase angles are practically the same for both cases since they depend on discharge flow conditions just outside the duct. (see Section 3.3).

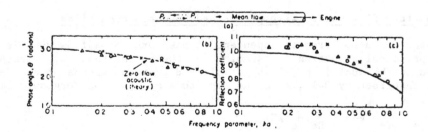

FIG. 3.4 Reflection of sound.(a) Exhaust tail pipe, $P_r = P_i Re^{i\theta}$
(b) Phase angle vs. frequency parameter. (c) Reflection coefficient vs. reflection parameter.——. Theoretical zero flow acoustic; o, M = 0.078; Δ, M = 0.11; x, M = 0.17.

Both examples show that with flow, the reflected wave amplitude exceeds that of the incident wave, for the lower wavenumbers. Though the pressure amplitude of the reflected wave has been increased, one can show, using equation 3.26, that the ratio of the reflected to incident intensity (energy flux) is equal to $(1-M)^2/(1+M)^2$, which is always less than unity.

(c)   Conservation relations for flow ducts

In developing the acoustic relations, with flow present, we must continue to conserve mass, energy and momentum and include the changes in pressure and velocity that are not directly related to the acoustic fluctuations but to the mean flow. For steady flow in a uniform (or slowly changing) duct we can assume to a good approximation that the flow is isentropic, so the stagnation enthalpy h_o remains constant. This condition is satisfied by

$$dh_o = o = dh + VdV \quad ,$$

$$= T.d\sigma + \frac{dp}{\rho} + VdV \quad , \qquad\qquad 3.16$$

with $p = \bar{p} + p'$ etc, where $\bar{p}$ is the time average value and p' the fluctuation. In 3.16, T is the temperature, h the enthalpy and σ the entropy. For steady flow we can neglect entropy fluctuations – note: this may not be realistic for acoustic waves of very high amplitude –, so do. The perturbation variables can now be replaced by the appropriate acoustic variables, viz:

$$p' = p_i + p_r$$
$$Z_s v' = p_i - p_r$$
$$\rho' = p'/c_o^2$$

Substitution in 3.16 yields

$$\frac{dp'}{\bar{p}+p} + (\bar{v} + v')dv' = o \quad ,$$

which to first order becomes

$$\frac{d(p_i+p_r)}{\rho_o} + \bar{v}. \frac{d(p_i-p_r)}{\rho_o c_o} = o \quad ,$$

where the subscript zero now refers to ambient conditions. This equation can be integrated to give, for the pressure amplitudes

$$\hat{p}_i(1+M) + \hat{p}_r(1-M) = \text{a constant} \quad , \qquad\qquad 3.17$$

which expreses conservation of energy.   Note that this reduces to 3.14 if M = o

Conservation of mass can be expressed as

$$dm = o = A(\rho dv + vd\rho) \qquad\qquad 3.18$$

or

$$A\left[(\bar{\rho} + \rho')dv' + (\bar{v} + v')d\rho'\right] = o \quad ,$$

which to first order, reduces to

$$\frac{A}{c_o}\left[(1+M)dp_i - (1-M)dp_r\right] = o \quad .$$

So that, after integrating this equation, continuity of volume velocity flow is expressed by

$$\frac{A}{c_o}\left[(1+M)p_i - (1-M)p_r\right] = \text{a constant} \quad , \qquad\qquad 3.19$$

which reduces to 3.15 if M = o

Describing the duct termination by 3.13, we find that, with flow, the total pressure at some point ℓ can be described by

$$\left[(1+M)\hat{p}_i + (1-M)p_r\right]_\ell e^{i\omega t} = \hat{p}_i\left[(1+M)e^{-ik^+\ell} + R(1-M)e^{i(k^-\ell+\phi)}\right]e^{i\omega t} \quad ,$$

$$3.20$$

which reduces to 3.7(a) when no flow is present.   An equivalent expressed to 3.20 describing volume velocity at ℓ, can be derived using equation 3.19.

(d)   Conditions at discontinuities

With flow present, it may no longer be realistic to assume that isentropic flow exists at discontinuities, particularly at rapid changes of cross sectional area (i.e. duct expansion, contractions and open ends). As well as the end effects noted earlier for zero flow, there will be additional energy losses in the flow, which must be included in the analysis.   In particular it is no longer realistic to assume that the entropy change dσ is zero.   The appropriate variations can be expressed as (3.8(a))

$$d\sigma = \sigma' = \frac{1}{\bar{\rho} \cdot T} \phi \quad ,$$

$$\rho' = \frac{p'}{c^2} - \frac{(\gamma - 1.}{c^2} \phi \quad ,$$

where γ is the ratio of the specific heats of the gas and ϕ is a compli- cated function representing irreversible exchanges (3.8(a)).   Defining

$$\delta = -(\gamma - 1)\phi \quad ,$$

and assuming that $\rho_2 = \rho_1$, where ρ_1 refers to conditions before the discontinuity and ρ_2 to conditions well downstream, conservation of energy across the area discontinuity is expressed by

$$\hat{p}_{2_i} (1+M_2) + \hat{p}_{2_r} (1-M_2) = p_{1_i}(1+M_1) + p_{1_r}(1-M_1) - \delta/(\gamma-1). \quad 3.21$$

Conservation of mass flow across the area discontinuity is satisfied if

$$A_2\left[\hat{p}_{2_i}(1+M_2)-\hat{p}_{2_r}(1-M_2)\right] = A_1\left[\hat{p}_{1_i}(1+M_1)-\hat{p}_{1_r}(1-M_1)+\delta M_1\right]. \quad 3.22$$

Another relationship is required to specify the conditions each side of the discontinuity.   We note further that the instantaneous momentum flux should also be conserved across the discontinuity.   The instantaneous momentum flux is given by $\rho v^2 A + A.p'$ where $A.p'$ is the force acting due to the pressure gradient.   Momentum is conserved if

$$d(\rho v^2 A) + Adp' = o,$$

or

$$d(\rho v^2 A) + D(p'.A) = p'dA = o . \quad\quad\quad 3.23$$

Integration of this equation across the discontinuity, and noting

$$\int_{2}^{1} pdA = p_2(A_1-A_2) \text{ yields}$$

$$\hat{p}_{2_i}\left[A_1 + A_2(2M_2 + M_2{}^2)\right] + \hat{p}_{2_r}\left[A_1 + A_2^2(M_2{}^2 - 2M_2)\right]$$

$$= \hat{p}_{1_i}\left[A_1(1 + M_1)^2\right] + \hat{p}_{1_r}\left[A_1(1 - M_1)^2\right] + \delta A_1 M_1{}^2 \quad . \qquad 3.24$$

This relation with 3.21 and 3.22 provide all the necessary conditions for describing one dimensional flow across discontinuities, since δ may be eliminated between them, giving two equations defining the components of p_2 in terms of p_1. Provided satisfactory allowances, where necessary, are made for adjustment for the phase changes at discontinuities, (i.e. by a length correction) these expressions provide adequate predictions for practical purposes which agree with observation. Examples of their application to acoustic energy transport calculations are given in references (3.5) and (3.8a).

Note that these examples assume isentropic flow at area contractions, which is not substantiated by later experiments with higher Mach number flows. Good agreement is however obtained with observation for the analysis set out above.

Finally, the mean acoustic energy flux per unit area, in a duct with flows, can be calculated from the general expression.

$$I = (1 + M^2)\overline{p'v'} + M\left|\frac{\overline{p'^2}}{\rho_o c_o} + \rho_o c_o \overline{v'^2}\right| \quad , \qquad 3.25$$

where overbars represent time averages (see ref. (3.9)). This corresponds to the zero flow intensity when $M = 0$.

With plane wave propagation, 3.25 can be simplified further, since $v' = p'/\rho_o c_o$. In this case we find for acoustic energy propagation in the direction of the flow that

$$I^+ = (1 + M)^2 \overline{p_2/\rho_o c_o} \quad . \qquad 3.26$$

The energy flux per unit area against the flow will be $I^- = (1-M)^2\overline{p_2/\rho_o c_o}$.

References

3.5 P.O.A.L. Davies and R.J. Alfredson 1971 'Design of silencers for internal combustion exhaust systems' Vibration and noise in motor vehicles. Proceedings I. Mech. E., London, Vol.156, 17 - 23.

E. Meyer and E.G. Neumann, Physical Acoustics, Chapter 11,
(Published by Academic Press).

R.J. Alfredson and P.O.A.L. Davies, 1970. The radiation of sound
from an engine exhaust.   Journal Sound and Vibration Vol.13,
pp 389 - 408.

R.J. Alfredson and P.O.A.L.Davies, 1971. Performance of exhaust
silencer components.   Journal Sound and Vibration Vol.14,pp 175-196.

P. Mungar and G.M. Gladwell, 1969.   Acoustic wave propagation in
a sheared fluid contained in a duct.   Journal Sound and Vibration
Vol.9, pp 28 - 48.

C.L. Morfey, 1971.   Sound generation and transmission in ducts with
flow.   Journal Sound and Vibration Vol.14, pp 37 - 55.

3.3 Sound generation by flow acoustic coupling

Some detailed discussion of flow noise source characteristics and
mechanisms has already been presented in Chapters 1 and 2. We now turn
to yet another group of sources which are of particular relevance to duct
flows.   The generation of pseudo sound, or nearfield pressure fluctuations,
by unsteady or turbulent flow has already been mentioned. We shall now
consider the case where these pressure fluctuations are intensified or
amplified in the presence of a sound field.   These sources are often
classified as a non-linear behaviour from the viewpoint of acoustic
analysis.

Whenever a flow leaves a sharp downstream facing edge it separates
and forms a thin shear layer or vortex sheet.   Such sheets, which involve
a rapid transverse change of streamwise flow velocity, are very unstable.
They quickly develop waves, which grow rapidly, until the sheet rolls up
to form a train of vortices.   Flow separations also occur at upstream
facing corners, but the flow may re-attach before an ordered set of
vortices forms.   This flow system provides a qualitative physical
explanation for the existence of the acoustic end effects that are found
at area discontinuities in pipes and ducts.

Since vortex sheets are very unstable, they respond to any small
disturbance.   Therefore the development of the waves in the sheet is
strongly influenced by sound waves.   Though not by themselves strong
radiators of sound, periodically generated vortices can excite resonators
very strongly.   This is the basic mechanism of many musical instruments,
though here, the acoustic field in the resonator provides the sound waves
which control the behaviour of the vortex sheet.   Discrete tones are also
generated when a vortex sheet is directed onto a sharp wedge.   Such sounds

are called edge tones and their intensity can be quite high when
amplified by feedback.

We shall consider first the way sound fields influence the development
of the vortices.   Next, consider the excitation of resonators by the
pseudo-sound field associated with the vortex motions and finally the
generation and characteristics of edge tones.   A detailed study and
analysis of the stability of vortex sheets was first set out by Rayleigh.
(3.10).   It is quite comprehensive and provides an adequate background
to this subject.

(a)   Acoustically synchronised vortex motion

There have been a number of observations and studies of the periodic
shedding of vortices synchronised by a sound field going back over a
century.   An example, which illustrates the process, is provided by the
synchronised vortex motion generated at the sudden expansion of the flow
in a pipe when an acoustic field is also present.   The observations of
the pressure waves are compared with predictions in Figure 3.5.

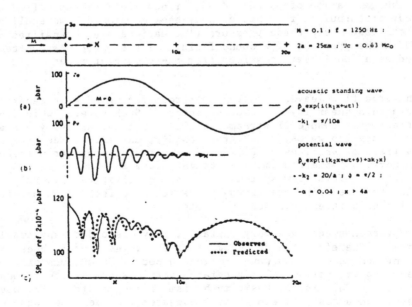

FIG. 3.5 An Acoustically Excited Duct Expansion

The peak SPL in the duct upstream was 130 dB, the frequency of the acoustic tone being deliberately chosen to match the length of the downstream duct to simplify later analysis.   The pressure measurements were obtained with a special probe microphone traversed along the duct axis.   The top diagram shows the acoustic standing wave in the duct with zero flow (Fig.3.5(a)). The amplitude and phase of the acoustic wave can be quite closely predicted by linear acoustic analysis.   The acoustic standing wave in the expansion can be described by

$$P_a(x,t) = \hat{p}_a \sin k_1 x e^{i\omega t} \ .$$

3.27

The form of the vortex travelling wave shown at (b) has been developed from a number of independent observations,[*] so it is a predicted estimate not a measurement.   The travelling wave is described by

$$P_v(x,t) = \hat{p}_v e^{-\alpha k_2 x} e^{i(\omega t - k_2 x - \phi)}$$

3.28

The combined wave p(x,t) is found by adding the two.   The mean square value was then calculated and has been compared with the SPL measurements in Figure 3.5(c).   The agreement is within the accuracy of the measurements.

At first the travelling wave motion, excited by the acoustic field in the vortex sheet, is strongly amplified.   But it then decays again further down the expanded duct.   Observations of the decay rate of pressure waves in flow ducts suggest that a common value of the attenuation coefficient α lies between 0.02 and 0.04.   The higher value seems appropriate for the vortex motion in this case.   The phase shift of $\pi/2$ corresponds to that for the build up of vibration amplitude in an undamped linear oscillator excited at resonance.   Finally, the wave number k_2 was calculated from the known phase velocity U_c of such vortex or shear wave motion. (Ref.(3.11)).

Though not very efficient radiators of sound in free space, vortex waves can excite resonators strongly, and they also generate a radiated dipole field when they induce fluctuating lift or drag forces on bodies immersed in the flow.   Both mechanisms represent significant sources of flow noise in flow duct systems and may provide an upper limit to sound attenuation by reactive silencer components.   The presence of the synchronous vortex pressure field also provides an explanation for the reflected wave amplitude p_r exceeding the incident amplitude p_i in Figures 3.3 and 3.4.
[*]   These observations are concerned with the initial development of a jet when forced by an acoustic field.   The data used here are unpublished but see Moore 1977,J.Fluid Mech 80,321-367 for similar information.

(b)   Resonators excited by vortices

Next consider the flow past a cavity or slot in the wall of a duct. The flow will separate at the upstream edge casting off a vortex sheet. In this case the sheet will be perturbed by eddies generated by the boundary layer growing along the wall upstream, as well as by any unsteady motions induced by flows into or out of the cavity. Once the sheet rolls up into a concentrated vortex it will induce a flow towards the cavity ahead of it and an outflow behind it as it passes over the cavity. The induced velocity field is illustrated in Figure 3.6 in relation to the vortices.

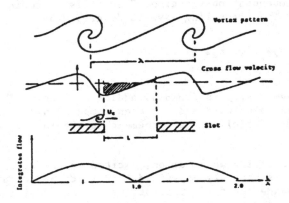

FIG. 3.6.   Inflow/Outflow Induced by Vortex Patterns

Excitation of the cavity depends on the amplitude of the integrated unsteady nett inflow (or outflow) to the cavity, which is a maximum when the vortex spacing λ is twice cavity length $L(L/\lambda = 0.5)$.   A maximum will also occur when $L/\lambda = 1.5$, 2.5 etc.   It has been observed that such cavities generate sound (3.12) with a Strouhal number given by

$$\frac{fL}{U_c} = (m-0.5); \quad m = 1, 2, 3, \text{ etc}, \qquad\qquad 3.29$$

where U_c is the phase velocity of the vortices.   If the Strouhal frequency corresponds to one of the natural acoustic inflow modes of the cavity, a resonance occurs and this then becomes a significant source of

sound.   A similar mechanism is involved in the generation of strong tones
by aircraft wheel wells, mentioned in Chapter 2.

The equation 3.29 applies to conditions when the convection speed is
$U_c << c_0$. At high flow speeds, (sonic or supersonic duct flows) one cannot
ignore the finite time taken by sound waves to travel across the slot from
down-stream to upstream edge.   A detailed quantitative analysis of such
flow problems can be found in Reference (3.13).

Flow excited resonance can provide significant acoustic sources in
lined ducts.   Such linings, consisting of a perforated sheet backed by
cells to form a continuous array of Helmholtz resonators, are fitted to
jet engine fan ducts for noise control.   Perforate sheet is also used
as a duct lining for similar purposes in exhaust pipes or ventillation
ducts.   In this case observations have demonstrated that cavities will
be excited at their Helmholtz frequency at the critical duct flow velocity
$V = fL$ (assuming $U_c \doteq V/2$, the average speed in the pipe boundary layer)
where f is the Helmholtz frequency.

(c)   Edge tones

If airflows from a slot onto a knife edge, edge tones are produced.
Experiment shows clearly that the pitch depends on the flow velocity,
the distance between the nozzle and knife edge and the presence of nearby
reflecting surfaces.   The mechanism involves the shedding of vortices from
alternate sides of the knife edge.   This produces an oscillating transverse induced velocity field at the slot, giving a transverse deflection
of the jet.   This oscillation maintains the alternate shedding of vortices
from the knife edge.   The Strouhal number is again found in accordance
with equation 3.29 with L = h, the distance of the knife edge or wedge
from the slot.   At least four modes or stages (1<m<4) of oscillation
have been observed experimentally by Brown (Ref.(3.14)) as indicated in
Figure 3.7.

The combination of an edge tone mechanism with a resonator provides
a very powerful source of radiation known as the Hartmann generator.
When driven by a choked nozzled they produce an intense radiation consisting of a series of harmonically related tones as shown in Figure 3.8

References

3.10   Lord Rayleigh 1896. The theory of sound, 2nd Edition, Chapter XX1,
        MacMillan (republished by Dover 1945).

3.11   M.J. Fisher and P.O.A.L. Davies 1964. Correlation measurements
        in a non-frozen pattern of turbulence. Journal Fluid Mechanics,
        Vol.18, pp 97.

3.12   S. Bolton 1976. The excitation of an acoustic resonator by pipe

3.13 J.E. Rossiter 1964. Wind tunnel experiments on the flow over
      rectangular cavities at subsonic or transonic speeds. Aero.
      Research Council, London.   R & M 3438.

3.14 G.B. Brown 1937. The vortex motions causing edge tones. Journal
      Physics Society, Vol.49, pp 493 - 521.

3.15 S.W. Coates and G.P. Blair 1974. Further studies of noise chara-
      cteristics of internal combustion engines. S.A.E. Trans. 83,
      740173.

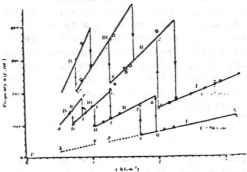

FIG. 3.7 Variation of Frequency with Reciprocal
         of wedge-distance for high velocities.

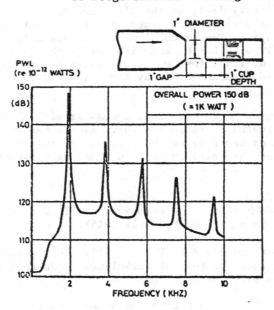

FIG. 3.8 Acoustic Output of 1 inch Harmann
          Generator

3.4   Source Identification and Characterisation

In section 3.1 above, we saw that acoustic energy propagation in ducts is by a wave mechanism, that can be described in terms of incident and reflected waves, characterised by a harmonic fluctuation in pressure and velocity.   We have seen also in 3.0 and 3.3 sources can be described in terms of cyclic discharges, unsteady flow, or by cyclic forces, or fluctuating pressure fields.   The ratio of the acoustic pressure to the velocity characterises the source impedance, which is often a complex quantity.   Finally, we saw in Figure 3.2 that the acoustic impedance of flow ducts can be very high (ideally infinite) at certain characteristic frequencies (or wavelengths), so that very high pressures are required to acoustically excite the duct.

As a first simple classification, we can relate unsteady discharges into (or out of) the duct to volume velocity sources and unsteady distribution of forces to pressure sources.   In particular, if the characteristics of the flow duct do not significantly modify conditions at the source, they can be characterised as constant volume velocity sources, as constant pressure sources, or a combination of both.   Where the source characteristics are modified by changes in the duct characteristics, then considerable testing would be required to describe such complex behaviour in a quantitative way.

There are examples, such as two stroke engines and ram compressors, where careful tuning of the inlet and exhaust duct is deliberately undertaken to optimise performance.   The approach is generally empirical, but a quantitative analysis is possible if the complete inlet, exhaust and cylinder system is analysed as a single combined unit.   Some progress has been made with this problem, employing the method of characteristics. It has not so far been shown that it is particularly suited to acoustic calculations for the prediction of noise (3.15).   The main reason for this is the amount of computation involved and the difficulty of specifying boundary conditions adequately for acoustic calculations.   We shall consider the application of linear acoustic theories to such problems in the next section.

To illustrate this point further, suppose we have a flow duct system with a source funtion $Q(t)$, a source impedance Z, a flow duct with an acoustic impedance Z_D and a termination impedance Z_w.   The incident pressure at the termination, p_t, will be given by

$$P_t = Q(t).Z_w/|Z+Z_D+Z_w|.\qquad\qquad 3.30$$

This says that the magnitude of the observed pressure at the termination depends on the source impedance Z and duct impedance Z_D, as

well as the termination impedance.   Measurements of P_t with known Z_w
and Z_D will provide the characteristics strength of the source $Q(t).Z$,
the product of its source function and source impedance.   For a flow
duct of constant area Z_D will vary with duct length as indicated in
Figure 3.2, while Z_w can be found from Figure 3.4.   If P_t is the incident
pressure at the termination, then, from 3.20 for a constant driving
pressure $Q(t)Z$ at the source.

$$\frac{P_t}{Q(t)Z} = G_p = \frac{1}{\left|(1+M)e^{-ik+\ell} + (1-M)re^{ik-\ell}\right|} \cdot \qquad 3.31$$

and for a constant driving velocity $V(t)$ at the source

$$\frac{P_t}{V(t)} = G_v = \frac{\rho_o c_o}{\left|(1+M)e^{-ik+\ell} - (1-M)re^{ik-\ell}\right|} \cdot \qquad 3.32$$

The function G_p and $G_v/\rho_o c_o$ are plotted in Figure 3.9.   It is clear that
a constant source will produce a measurement of P_t that varies by over
20 dB depending on frequency.

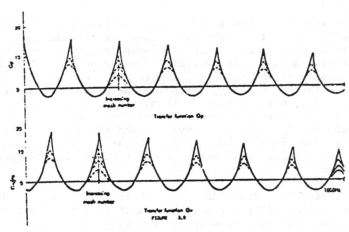

FIG.   3.9 Transfer Function G_v

If an acoustic filter, impedance Z_f, is inserted in the duct,
schematically the situation becomes as sketched in the figure below

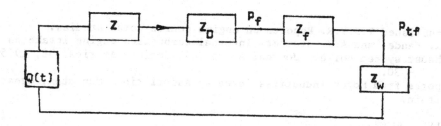

The incident pressure before the silencer will be

$$P_f = Q(t)(Z_f+Z_W)/|Z+Z_D+Z_f+Z_W| \ , \qquad\qquad 3.33$$

and the new duct termination incident pressure

$$P_{tf} = Q(t)Z_W/|Z+Z_D+Z_f+Z_W| \ . \qquad\qquad 3.34$$

Assuming the forcing function $Q(t)$ is unaltered, then the change in transmitted sound is P_{tf}/P_t, this is called the insertion loss of the modified
system.   Many silencer designs consider only the transmission loss of
the system which puts $Z = 0$ in 3.30, 3.33, and 3.34.   Or they assume
that p_f is constant. so that the change in sound transmission is
$Z_W/(Z_f+Z_W)$, which is not realistic.

     In many practical cases it is impossible to predict the source
function or source impedance from first principles and they must be
ascertained by direct measurement.   Adequate specification of a complex
source will involve measurements with at least two configurations.
Exhaust noise source characteristics of a high output diesel engine
produces one such example.   This includes sound generation by pulsatile
flow, valve noise, turbine noise and by flow-acoustic interaction.
Such measurements include the further problem that the exhaust gases
are hot and so temperature gradients normally exist along the duct with
corresponding changes in the value of the speed of sound and the specific
acoustic impedance ρ_c.

References

I   I.C. ENGINE SOURCES
(a)   M.L. Munjal and A.V. Sreenath, Department of Internal Combustion
      Engineering, Indian Institute of Science, Bangalore, India.
      (Various internal and external reports).

(b)   Unpublished I.S.V.R. Reports on engine source noise studies.
(c)   E.K. Bender and A.J. Brammer. Internal combustion engine intake and
      exhaust system noise. Journal Acoustical Society America, 58, 1975.
      pp 22 - 30.
(d)   Reports from Motor Industries Research Association, Nuneaton, Great
      Britain.

II   VALVE SOURCES
Valve manufacturers' data and performance sheets.

III   FANS AND BLOWERS
Fan and blower manufacturers' data and performance sheets.

3.5   Duct Propagation of High Amplitude Pressure Waves

        In developing acoustic analysis, the system of equations is first
linearised by assuming that pressure and density perturbations are so small
compared with ambient values, that their products are second order and
may be neglected.   Similarly squares or products of acoustic particle
velocity with other fluctuating quantities may similarly be neglected.
Pressure waves of high intensity exist in a number of practical flow duct
problems, so one is concerned to identify those cases where departures
from linear acoustic theory are sufficiently serious to warrant an alternative approach.

        For plane waves of finite amplitude, there is a simple exact solution due
to Riemann $[3.16\ Pg.40]$. He showed that the one-dimensional isentropic flow
equations have two particular solutions in which all flow variables
(p, ,v etc.) are propagated at the same velocity in the x direction.
This velocity, known as the signal velocity, u, is given by

$$(dx/dt)^+ = u^+ = (v+c)\ (+\text{x travelling waves})$$
$$(dx/dt)^- = u^- = (v-c)\ (-\text{x travelling waves})$$

                                                                          3.35

For each solution there is a unique velocity-pressure relationship which
does not depend on waveform.   The velocity is given at constant entropy
by

$$v = \int_{P_o} (\rho c)^{-1} d\rho \quad (+\text{x travelling waves })$$

                                                                          3.36

$$v = -\int_{P_o} (\rho c)^{-1} d\rho \quad (-\text{x travelling waves})$$

One can develop an analytical approach based on these two equations, to describe the +x and −x travelling waves, called the method of characteristics. (See Landau and Lifschitz (3.17)).

These waves, in the limit of small perturbations, correspond to acoustic waves discussed hitherto. They differ in two main respects.

(i)   The rule of superposition does not apply.

(ii) With waves of finite amplitude, the signal velocity u is a flow variable, so that different values of pressure travel at different speeds. +ve pressures travel at speeds above the acoustic velocity c_0. The result is that the waveform changes (distorts) as the wave travels.

In a perfect gas we find that

$$u^+ = (c_0 + \frac{\gamma - 1}{2} v) \qquad\qquad\qquad 3.37$$

where γ is the ratio of the specific heats.

For moderate amplitudes, where terms like p' can be neglected, the excess wave velocity, $u' = u^+ - c_0$, is given by (γ = const)

$$u' = \frac{\gamma + 1}{2} \frac{p'}{\rho_0 c_0} . \qquad\qquad\qquad 3.38$$

Consider a frame of reference c_0 moving in the x direction. Relative to this frame each point on the wave moves at a velocity u' so a +ve pressure wave steepens and a negative wave flattens out. The wave steepening can only proceed until the pressure rise become instantaneous when a shock front or shock wave forms.

A shock front will develop after a time t_* given by

$$t_* = \frac{c}{\frac{1}{2}(\gamma + 1)} \left| - (\frac{\partial x'}{\partial p'})i \min \right| \text{(isentropic).}$$

where the subscript i refers to the initial waveform at t = o. It is not adequate to neglect viscous effects while shock waves involve irreversable changes, so the isentropic flow assumptions are hardly valid. The analysis is involved, but two approaches can be found in either 3.18, or 3.19. It is of some practical importance to know the range and frequency over which sound waves can be treated as linear. Figure 3.10 provides

a partial answer for plane waves in air, where the attenuation is given as
a function of temperature and water vapour content. With levels of
15-20 dB above the linear behaviour line significant steepening of the
waveform will occur.

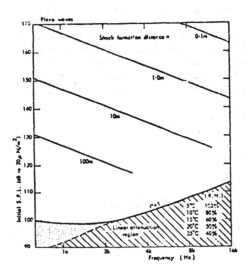

FIGURE 3.10

On the other hand, many practical flow ducts have relatively short
runs between discontinuities, so that wave steepening may not be complete
before the wave is modified. One should not overestimate the practical
effects of wave steepening. It is worth noting for instance, that 3 dB
represents a factor of 2 in terms of acoustic energy. Linear acoustic
models may be used to predict energy transport to within closer limits
than this, with pressure pulses of up to 0.5 bar or more in amplitude.

References

3.16 B. Riemann, 1860 Abhandlungen Ges. Wiss. Göttingen 8, pp 43 - 65

3.17 L.D. Landau & E.M. Lifshitz 1959. Fluid Mechanics. Pergamon Press,
      Oxford.

3.18   M.J. Lighthill 1956. <u>See</u>: Surveys in Mechanics (ed. G.K. Batchelor and R.M. Davies), Cambridge University Press, London.

3.19   R.T. Beyer 1965. Physical Acoustics, Chapter 7 (ed. W.P. Mason), Vol. IIB, Academic Press, New York.

3.6   The radiation of sound from duct walls

There are many practical noise control problems that require an accurate prediction of noise climate at the design stage. A major contribution is noise from high pressure control valves and reaction vessels in refineries and process plants, from pipelines or exhaust ducts from large compressors or diesel engines and from ductwork of ventillating or extracting systems.   They all exhibit the common feature of noise radiation from bounding surfaces excited by internal pressure fields.

The problems of noise climate prediction in the vicinity of such ducts requires a specification of the internal noise climate, and an understanding of the response of the duct or vessel to the applied field. The sources, or the internal pressure field, includes both the propagating acoustic fields as well as the non-propagating fields that are generated at duct discontinuities as well as the pseudo-sound fields associated with boundary layer flows and vortex shedding.   The noise climate predictions also require some quantitative understanding of the radiated sound field from distributed sources.

Where the excitation is purely acoustic, or arises from propagating sound waves, the internal pressure field can be predicted by the various methods outlined earlier in this course.   Basic understanding, or experience, remains inadequate at present, to include those cases where high order mode wave generation or propagation, or flow generated pressure fields, provide significant wall excitation, though there has been some recent progress in this field.(3.20).   Finite element methods may be appropriate here.

The response of the pipe or duct system (including bends and discontinuities) can be calculated by a statistical approach (3.21) or by a finite element approach (see previous lectures).   The effort involved in obtaining such detailed solutions may not be warranted for many practical prediction purposes.   The aim of some current fundamental research in this field is to provide basic physical understanding and sufficient data to set up realistic noise prediction procedures and codes of practice for such problems.

The majority of practical problems involve the acoustic excitation of a pipe or duct by unsteady or turbulent flow through valves of various types.   Prediction of the noise radiation from the duct surface requires a comprehensive specification of the internal noise climate by spectral distribution of the pressure field as well as its overall sound power

level.   The spectrum of valve noise depends on a number of factors
including the dimensions of the valve, the geometry of the orifice,
body and connecting pipe, the effective acoustic impedance of source
and duct with coupled resonances between the flow and system geometry.

There are, however, among the data, indications that different
mechanisms are related to different parts of the noise spectrum.   Thus
higher frequency components are related to jet noise and intermediate or
lower frequency components are related to local acoustic resonances or
flow-acoustic interaction.   It seems clear that a satisfactory prediction
of valve source noise from theoretical principles alone is unlikely.
Thus experimental methods seem essential, though such data must be
appropriately scaled for practical application so a basic understanding
of the mechanisms of noise generation is also necessary.

3.6.1   Prediction of valve noise

An example is provided by recent studies to provide practical noise
prediction procedures for gas regulator valve noise (3.22).   The procedure
is based on finding the answer to two questions (a) specification of the
acoustic power generated by the valve and (b) specification of the acostic
transmission loss (TL) of the pipe wall.   Attenuation of the sound
travelling along the pipe and reflection of sound from junctions, and
branches, must also be taken into consideration.

Valve noise data is now provided by valve manufacturers but is
generally limited to data or noise radiation from the pipe wall close to
the valve.   This does not provide the information necessary or predictions
of noise radiation when the transmission loss of the pipe wall differs
from their test conditions, when pipe runs are long, or when bends or duct
junctions are significant factors.

The scheme involves measurement of the valve sound power spectrum
operated at a systematic series of pressure ratios and percentage openings.
This datais analysed and stored in a data bank.   The noise signature of
the valve for any specified operating condition can then be recovered by
four-point interpolation from the data bank.   The results are scaled for
changes in valve size, pipeline pressures or gas physical properties.

Pipe wall transmission loss was measured from a series of pipe wall
noise reduction measurements.   The loss depends roughly on the cube of
the wall thickness to pipe radius ratio h/r and several other factors.
It can be expressed as

$$TL = -10\log_{10} \left| \frac{8\pi^2 \, \rho_{in}\rho_{ex} \, c_{in}^3 \, c_{ex} \, n(\omega)\sigma_{in} \, \sigma_{ex}}{\omega^3 \, W^3 A\eta} \right| \qquad 3.39$$

where ρ and c are gas densities and sound speeds.   Also n(ω) is the pipe wall model density, σ the radiation efficiency, W the area density, η the internal loss coefficient and A the pipe wall surface area, with ω the radian frequency.

The value of radiation efficiency rises to unity at the critical wave coincidence frequency, and remains at unity for higher frequencies. At high frequencies $n(\omega) \propto h^{-1}$ with $W \propto h$, so acoustic transmission above coincidence is $\propto h^{-3}$.   Below coincidence, the variation of σ with h makes transmission $\propto h^{-1}$.   We see also that internal pressure (ρ_{in}) modifies the TL.   The results scale with larger pipes provided the ratio h/r is held constant, with a corresponding shift in frequency scales.   The pipe wall internal loss coefficient η is assumed to vary as $(\omega)^{-\frac{1}{2}}$.

Other factors which affect prediction are the attenuation of the field in the pipe, with distance from the source, and any increase in local pressures by wave reflections at discontinuities.   Some typical measurements of wall noise reduction are given in Figure 3.11.   The variation with frequency underlines the need for specific evaluation of the spectral distribution of the pressure field, for prediction purposes. The predicted radiated sound field from the pipe downstream of a 200mm valve is compared with measurements in Figure 3.12.   The predictions were based on valve noise measurements on a 50mm valve of similar type and construction, being suitably scaled to provide the predictions,   It can be seen that the agreement is reasonably good.

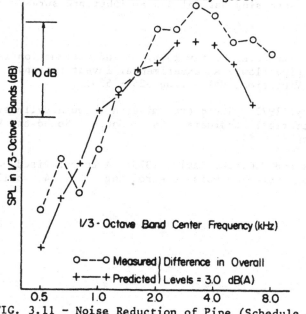

FIG. 3.11 - Noise Reduction of Pipe (Schedule 40)

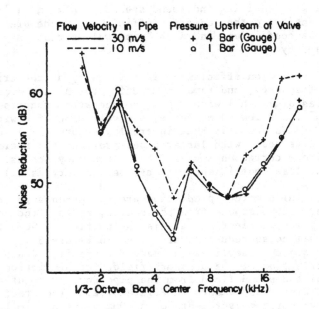

FIG. 3.12 – Measured and Predicted Spectra of 200mm valve
Assessing Natural Gas at 15bar pressures

References

3.20 G.F. Kuhn   and C.L. Morfey 1976.   Sound attenuation in fully
      developed pipe flow – an experimental investigation.   Journal
      Sound and Vibration, Vol.44, pp 525 – 529.

3.21 E. Szechenyi 1971.   Sound transmission through cylinder walls
      using statistical considerations.   Journal Sound and Vibration,
      Vol. 19, pp 83 – 94.

3.22 D.J. Small and P.O.A.L. Davies 1975.   A computerised valve noise
      prediction system.   Noise Control Eng. Vol. 4, pp 124 – 128.

APPENDIX A

ACOUSTIC ANALOGIES FOR SOURCES OF AERODYNAMIC SOUND

A1.0   Lighthill's acoustic analogy

In considering the sound generated by fluid motion it is clear that a practical and empirical understanding of the relation between moving fluids and the generation of melodic and other sounds must have existed from earliest times.   Many musical instruments employ a device for producing a periodically separated flow with a resonator for sound amplification.   Many of such acoustic source mechanisms are qualitatively well understood, at least.   They involve the periodic separation of the flow at some edge or surface to produce a discontinuity that can be described mathematically by a sheet of vorticity.   Such sources are known generally as edge tones.   Aerodynamic sound is also generated by an air flow in the absence of any solid boundaries.

An exact hydrodynamic analysis of the production of sound by a turbulent flow was developed by Lighthill (1), (2) by combining the equations describing the conservation of mass and of momentum, for an elementary fluid volume.   In the absence of mass injection and externally applied forces, they are respectively

$$\frac{\partial \rho}{\partial t} + \frac{\partial}{\partial x_i} (\rho v_i) = 0 \qquad\qquad A(1)$$

and

$$\frac{\partial}{\partial t} (\rho v_i) + \frac{\partial}{\partial x_j} (\rho v_i v_j) + \frac{\partial \sigma_{ij}}{\partial x_j} = 0 \qquad\qquad A(2)$$

where σ_{ij} is the viscous stress tensor.

The equations are written in tensor notation, implying summation over repeated suffixes, while i, and j, take the values 1,2,3.   Thus, written out in full equation (1) becomes

$$\frac{\partial \rho}{\partial t_1} + \frac{\partial}{\partial x_1} (\rho v_1) + \frac{\partial}{\partial x_2} (\sigma v_2) + \frac{\partial}{\partial x_3} (\rho v_3) = 0,$$

where x_1, x_2, x_3 are three orthogonal co-ordinates, with v_1, v_2 and v_3 the corresponding components of the velocity vector.

Lighthill eliminated the mass flux ρv_i from equation (1) and (2), by subtracting the space derivation of (2) from the time derivation of (1) to yield

$$\frac{\partial^2 \rho}{\partial t^2} = \frac{\partial^2}{\partial x_i \partial x_j} (\rho v_i v_j) + \frac{\partial^2 p}{\partial x_i^2} + \frac{\partial^2 \sigma_{ij}}{\partial x_i \partial x_j} \quad . \qquad\qquad A(3)$$

He then converted this to a wave equation, by subtracting the term $c_o^2 \partial^2 \rho / \partial x_i^2$ from both sides to give

$$\frac{\partial^2 \rho}{\partial t^2} - c_o^2 \frac{\partial^2 \rho}{\partial x_i^2} = \frac{\partial^2}{\partial x_i \partial x_j} \left[\rho v_i v_j + (p - c_o^2) \delta_{ij} - \sigma_{ij} \, , \right] \qquad\qquad A(4)$$

where $\delta_{ij} = 1$ if $i = j$ and $\delta_{ij} = 0$ if $i \neq j$.

The term in square brackets on the right-hand side of (4) is known as the Lighthill stress tensor T_{ij}. If T_{ij} is zero, then (4) describes the propagation of density in the manner of sound waves. In fact, outside the flow region where all but the acoustic velocities are zero, T_{ij} will be zero, at least to the accuracy of linear inviscid acoustic theory.

Since in an acoustic wave the particle velocities are small, products like $v_i v_j$ are second order and may be neglected. Similarly, since the pressure changes are very nearly isentropic in an acoustic wave, the term $p - c_o^2 \rho$ is also of second order. This leaves only the viscous term in T_{ij} which represents the attenuation of sound waves, which is normally only significant over relatively large distances of propagation. We note further that, for acoustic waves propagating in a homogeneous medium at rest, the inviscid mementum equation (2) is to the same approximation (first order) given by

$$\frac{\partial}{\partial t}(\rho v_i) + c_o^2 \frac{\partial \rho}{\partial x_i} = 0 \qquad\qquad A(5)$$

since $\partial p / \partial x_i = c_o^2 \partial \rho / \partial x_i$.

Thus the tensor T_{ij} is only effectively non-zero inside the flow, while outside the flow the density fluctuations propagate like sound waves. Thus equation (4) may be rewritten as

$$\frac{\partial^2 \rho}{\partial t^2} - c_o^2 \frac{\partial^2 \rho}{\partial x_i^2} = \frac{\partial^2 T_{ij}}{\partial x_i \partial x_i} \quad , \qquad\qquad A(6)$$

where the right-hand side is regarded as an acoustic source.

For an observer at $\tilde{x}$ outside the turbulent flow region at $\tilde{y}$, the solution to (5) is

$$\rho(x,t)-\rho_o = \frac{1}{4\pi c_o^2} \int_V \frac{\partial^2}{\partial y_i \partial y_j} T_{ij}(\tilde{y},t - \frac{|\tilde{x}-\tilde{y}|}{c_o}) \frac{dV(\tilde{y})}{|\tilde{x}-\tilde{y}|} \quad . \qquad A(7)$$

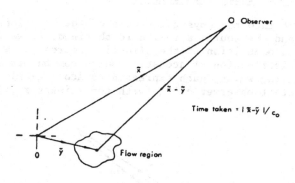

FIG. 1.

FIGURE 1.

This equation gives the result in terms of a fluctuating density ρ, but if desired, can be converted to pressure p by putting $p=c_o^2\rho$, which will apply in most practical cases, to the same approximation as (5).

This result shows that the density fluctuations at $\tilde{x}$ are those that would be produced by an array of sources of strength $\partial^2 T_{ij}/\partial y_i \partial y_j$ per unit volume, distributed over the flow region V. In the integration, due account has been taken of the fact that the sound, arriving at $\tilde{x}$ at time t, must have been emitted at an earlier time $t' = t - |\tilde{x}-\tilde{y}|/c_o$. The interval will vary with the component source position $\tilde{y}$. The sources in (7) are termed <u>acoustically equivalent</u>, meaning that they produce the radiated field at $\tilde{x}$ equivalent to that due to the flow, but for the solution, these sources are used to replace the flow in an acoustic medium that is now at rest everywhere.

The physical nature of the equivalent sources can be determined from equation (4). First there are the stresses acting in the turbulent fluid. The dominant stress term, in the absence of solid boundaries, is generally $\rho v_i v_j$, which represents the transport of components of fluid momentum (per unit volume) ρv_i by velocity components v_j. Such stresses arise from momentum exchanges associated with turbulent mixing in the flow and are widely known as the Reynolds' stresses. The other stresses are those

associated with the fluctuating pressures $p\delta_{ij}$ and the vicous stresses
σ_{ij}. Finally there is the stress which would be created by the density
perturbation, contained in the $c_o^2\rho$ term. Thus the effective source
strength is composed of the resultant of the difference between the
effective stresses acting on the fluid and those imposed by the density
field.

A1.1   Wave models of distributed sources

Further insight into the physical nature of the relation between the
residual stresses and the source strength is obtained, if we first
perform a Fourier decomposition of the effective sources. We then can
suppose we have a distribution of acoustic source components that can be
described by travelling waves, each expressed as $A\cos(\omega t - kx)$. Suppose
also there is a distant observer at r_0 (with co-ordinate r_0, θ), as shown
in Figure 2.

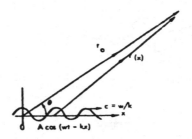

FIG. 2.

Components of the sound will arrive at times $t - r(x)/c$ after they are
emitted. If the observer is remote, so the angle θ is effectively
constant, then

$$\tilde{r}(x) = \tilde{r}_o - \tilde{x}\cos\theta.$$

Thus the apparent source strength, seen by the observer at $\tilde{r}_o$, will be
given by

$$\int_{-\infty}^{\infty} \frac{A}{r}\cos\left|\omega(t - \frac{r_o}{c_o} + \frac{x\cos\theta}{c_o} - kx\right| dx$$

and setting $t' = t - {}^r o/c_o$, with $\omega = ck$ this becomes

$$\int_{-\infty}^{\infty} \frac{A}{r}\cos\left|\omega t' + (\frac{c\cos\theta}{c_o} - 1)kx\right| dx .$$

As $r_o \to \infty$, this integral is made up of cosine and sine terms of the form

$$\frac{A}{r_o} \cos \omega t' \int_{-\infty}^{\infty} \cos \left| (\frac{c \cos\theta}{c_o} - 1)kx \right| dx .$$

The only non-zero value for all the integrals is when $|c \cos\theta/c_o - 1|$ is zero, that is if each component phase velocity satisfies

$$c = \frac{c_o}{\cos\theta} . \qquad\qquad A(8)$$

This result shows that only those wave number components of a distributed travelling source field, that possess supersonic phase velocities, make any contribution to the far field. This fact can be used to identify that part of the stress tensor T_{ij} that must be quantified to estimate the radiated noise. One can note, in addition, that supersonic phase velocities can only exist for components of a spectral representation of a flow pattern, whose structure is changing relatively rapidly.

A1.2   Classical models of acoustic sources (after Lighthill)

(a) The monopole

The simplest representation of an acoustic source is to suppose that, at some point $\tilde{y}$, there is a small fluctuating source of fluid $q(t)$. At this point conservation of mass is represented by

$$\frac{\partial\rho}{\partial t} + \frac{\partial}{\partial x_i} (\rho v_i) = q(t) . \qquad\qquad A(9)$$

Elimination of (ρv_i) between this equation and (5) by the process adopted earlier yields the wave equation

$$\frac{\partial^2\rho}{\partial t^2} - c_o^2 \frac{\partial^2\rho}{\partial x_i^2} = \frac{\partial}{\partial t} (q) . \qquad\qquad A(10)$$

Comparison of this equation with (6) illustrates the adoption of the description acoustically equivalent for the sources in (7). For an observer at $\tilde{x}$, the solution for (10) is

$$\rho(\tilde{x},t) - \rho_o = \frac{1}{4\pi c_o^2 |\tilde{x}-\tilde{y}|} \frac{\partial}{\partial t} q(t - \frac{\tilde{x}-\tilde{y}}{c_o}) . \qquad\qquad A(11)$$

It is not difficult to show that this result can be extended to any combination of point sources, distributed over a region of space. Their contributions would then add linearly, with appropriate allowances for relative phase and retarded time, to provide the total acoustic field experienced by a remote observer. Acoustic source distributions of higher order can be represented by appropriate combinations of elementary sources or monopoles which may be so arranged that their individual contributions almost cancel in some specified directions of propagation, and are reinforced in others.

(b) The dipole

Suppose now that at the point $\bar{y}$ there is a local fluctuating force field $f_i(y,t)$. This would be represented by an equal source and sink in close proximity, a combination in hydrodynamics known as a doublet or dipole. In this case conservation of momentum would be described by (5) except in the region at $\bar{y}$, where the term $f_i(y,t)$ must be added to the right-hand side. Eliminating (ρv_i) between equations (1) and (5) gives the wave equation

$$\frac{\partial^2 \rho}{\partial t^2} - c_o^2 \frac{\partial^2 \rho}{\partial x_i^2} = -\frac{\partial f_i}{\partial x_i} \qquad\qquad A(12)$$

One can show that such a point force will radiate sound in a directional manner. If the angle between the force vector and the direction vector of the observer is θ then the magnitude of the resultant field is proportional to $\cos \theta$.

Consider now a moving body in a fluid which produces a force distribution F_i over a volume $V(\bar{y})$, for this case F_i will replace f_i in (12). The result, for an observer at $\bar{x}$, with $t' = t - (\bar{x}-\bar{y})/c_o$, will be

$$\rho(\bar{x},t) - \rho_o = \frac{-1}{4\pi c_o^2} \int_V \frac{\partial F_i}{\partial y_i} \frac{(y,t')}{(\bar{x}-\bar{y})} \, dv(y) \quad . \qquad\qquad A(13)$$

To evaluate this result in practice we would have to calculate the contribution for each element of the source region V at its appropriate delayed time.

It is often more convenient to transform the space derivative co-ordinates from the source y_i to the observer x_i. To do this, we first set $\tilde{r} = |\bar{x}_i - \bar{y}_i|$. Further using Gauss' theorem, since F_i is zero outside the source region, we know that

$$\int_V \frac{\partial}{\partial \bar{y}_i} \left(\frac{1}{\tilde{r}} F_i\right) dV = \int_S \frac{1}{\tilde{r}} F_i dS_i = 0.$$

so that we can expand the volume integral as

$$0 = \int_V \frac{1}{r} \frac{\partial F_i}{\partial \tilde{r}_i} \, dv + \int F_i \frac{\partial}{\partial \tilde{y}_i} \left(\frac{1}{\tilde{r}}\right) dv \ .$$

Also, since

$$\partial r / \partial x_i = -\partial r / \partial y_i \text{ and } F_i \text{ depends on } y \text{ but not on } x, \text{ we}$$

find that

$$\int_V \frac{1}{r} \frac{\partial F_i}{\partial y} \, dv = \frac{\partial}{\partial x_i} \int_V \frac{1}{r} F_i \, dv.$$

Thus equation (13) may also be written as

$$\rho(\tilde{x},t) - \rho_o = - \frac{1}{4\pi c_o^2} \frac{\partial}{\partial x_i} \int_V \frac{F_i}{r} \left(y, t - \frac{r}{c_o}\right) dv(y). \qquad A(14)$$

This appears similar to (13), but is really very different. In (13), the equivalent sources are a set of almost cancelling monopoles of effective strength proportional to the space derivative of F_i. In (14) the source is a dipole, but the whole force distribution is acoustically equivalent to a single dipole, of strength given by the integral in (14), with higher order contributions which may arise from the details of the distribution.

(c) The quadrupole

One can arrange two adjacent similar point dipoles (or the equivalent four monopoles) in two ways, so that the resultant represents a momentum fluctuation at a point. This provides an acoustic source of higher order, called a quadrupole. It can be shown that one arrangement with the two cancelling dipoles arranged side by side, will produce a directional field, given by $\cos^2\theta$, with maxima in four directions, like a clover leaf. The other arrangement, with the forces in line, will give something like a $\cos\theta\sin\theta$ type of directional distribution.

A compact region of turbulence will radiate sound like an acoustic quadrupole (equation (6)). Again for a distribution of quadrupole sources represented by the integral in (7), evaluation of the integral over a flow region V can be simplified, by repeating the manipulation employed to derive (14) from (13). Doing so provides the more convenient form

$$\rho(\tilde{x},t) - \rho_o = \frac{1}{4\pi c_o^2} \frac{\partial}{\partial x_i \partial x_j} \int \frac{1}{r} T_{ij}(\tilde{y},t')dv(y) \qquad A(15)$$

In (15), provided the source region is acoustically compact, the flow is now acoustically equivalent to a single quadrupole, of total strength given by the volume integral.

Finally, in a formal sense, one could consider that a combination of sources of different order can always be represented by an equivalent scaler source $Q(t)$. It is physically more useful, generally, to retain the dipole or quadrupole formulation for a detailed analysis of aerodynamic sound sources.

A1.3   Aerodynamic sound radiation to a remote observer

For an observer at a relatively large distance from the source region, we are primarily concerned with the far-field radiation. Thus terms which decay with distance at a rate greater than $1/r$ may be neglected. To calculate the far-field radiation at a distant observer, produced by a fluctuating force distribution at y, we first rewrite (14) to take account of this change as

$$\rho(x,t) - \rho_o = \frac{-1}{4\pi c_o^2} \int_V \frac{\partial r}{\partial x_i} \frac{\partial}{\partial r} \left| \frac{F_i(y,t')}{r} \right| dv(y) \ .$$

Also, since $r^2 = (x_i - y_i)^2$, $\partial r/\partial x_i = (x_i - y_i)/r$, while $\partial F_i/\partial r = -1/c_o(\partial F_i/\partial t)$. Retaining only terms decaying at $1/r$, this equation may be rewritten as

$$\rho(x,t) - \rho_o = \frac{1}{4\pi c_o^3} \int_V \frac{x_i - y_i}{r} \cdot \frac{1}{r} \frac{\partial F_i}{\partial t} (y,t')dv(y) \ . \qquad A(16)$$

For a single point force f_i, the term $(x_i - y_i)/r$ is equivalent to $\cos\theta$ and relative to the i direction thus represents the directional distribution The same term in (16) represents the resultant directional distribution of the resultant single dipole.

Comparing (16) with (11) we see also that there is an extra factor of $1/c_o$ in the relative strength of the field from a point dipole compared with that from a monopole of similar source strength. Thus a point dipole is acoustically less efficient as a far-field radiator of sound than a monopole.

By a similar manipulation one can also show that the far-field radiation from a quadrupole source region can be expressed as

$$\rho(\tilde{x},t) - \rho_o = \frac{1}{4\pi c_o^4} \int_V \frac{(x_i-y_i)(x_j-y_j)}{r^2} \cdot \frac{1}{r} \frac{\partial^2 T_{ij}}{\partial t^2}(\tilde{y},t') dv(\tilde{y}). \qquad A(17)$$

Here the term $(x_i-y_i)(x_j-y_j)/r^2$ describes the directional distribution of the radiated sound field. Again we note that there is a further factor of $1/c_o$ in the relative strength of the field from a point quadrupole compared with that from an equivalent point dipole. Thus a quadrupole is acoustically less efficient as a far-field radiator of sound than a dipole.

The final step is to assume that the source region is small compared with the distance to the observer so that $(\tilde{x}) >> (\tilde{y})$ so that (17) can be simplified to

$$\rho(\tilde{x},t) - \rho_o = \frac{1}{4\pi c_o^4} \frac{x_i x_j}{r^3} \int_V \frac{\partial^2 T_{ij}}{\partial t^2}(\tilde{y},t') dv(y) \qquad A(18)$$

with an equivalent expression in place of (16). It may be more physically meaningful to combine the direction cosines $x_i x_j/r^2$ with T_{ij} to form a scaler source strength. Neglecting the viscous term in T_{ij}, the source strength becomes

$$\rho \frac{v_i x_i}{r} \frac{v_j x_j}{r} + (p - c_o^2 \rho) \delta_{ij} \frac{x_i x_j}{r^2} .$$

Now $u_i x_i/r$ is simply the resolved part of the velocity in the direction of the observer, which we denote by u_r, while $\delta_{ij}x_i x_j/r^2 = x_i x_i/r^2 = 1$. Thus the quadrupole source strength can be expressed as

$$T_{rr} = \rho u_r^2 + p - a_o^2 \rho . \qquad A(19)$$

In a sound field $p - p_o = c_o^2(\rho-\rho_o)$, so the far-field pressure amplitude can be expressed as

$$p(\tilde{x},t) - p_o = \frac{1}{4\pi c_o^2 r} \int \frac{\partial T_{rr}}{\partial t^2}(\tilde{y},t') dv(y) . \qquad A(20)$$

A similar substitution may be performed in the case of other types of acoustic source function.

A1.4   Sources in motion

When sources are in motion it is often more convenient to perform the integrations in co-ordinates η, related to the moving source.   If the sources are moving with a velocity u, corresponding to a Mach number $M=u/c_o$, the $\tilde{\eta}$ and $\tilde{y}$ co-ordinate systems are connected via

$$\tilde{\eta} = \tilde{y} - \tilde{M}c_o t.$$

However, in performing the integrations over the source region we need to calculate the appropriate time $t - r/a_o$, or over

$$\tilde{\eta} = \tilde{y} + \tilde{M}r - \tilde{M}c_o t .$$

Thus the co-ordinate transformation to moving from fixed axis is appropriately performed by

$$\tilde{\eta} = \tilde{y} + \tilde{M}r$$

as first suggested by Lighthill.

The tranformation also modifies the volume element of the integration, in that $d(v)y$ is replaced by $d(v)\eta/(1-M_r)$ where $M_r = M_i(x_i-y_i)/r$. Similarly, transformation from $\partial r/\partial x(y)$ to $\partial r/\partial x(\eta)$ will involve a further factor $(1-M_r)^{-1}$.   Thus the noise experienced at r from a force field moving at a Mach number M_i can be expressed, in co-ordinates moving with the source, by transforming (16) to

$$p(x,t) - p_o = \frac{1}{4\pi c_o} \int \frac{x_i-y_i}{r(1-M_r)} \; \frac{1}{r} \frac{\partial}{\partial t} \frac{F_i(\eta,t')}{(1-M_r)} \; dv(\eta) \qquad A(21)$$

We see, for example, if the sources are moving at a velocity $\bar{u}$ which makes an angle θ with the direction of the observer, then the apparent strength of a moving monopole is increased by the doppler factor $1/(1-M\cos\theta)$.   Similarly, the apparent strengths of moving dipoles and quadrupoles are increased by the factors $1/(1-M\cos\theta)^2$ and $1/(1-M\cos\theta)^3$ respectively, when they are evaluated in terms of co-ordinates moving with the source.

A1.5   Turbulence as a source of sound

Turbulent flow is the most common source of fluid motion and may be regarded as a complex assembly of locally organised but unsteady velocity patterns, which interact strongly with each other as they move with the flow. We are concerned here with a description of the characteristics of turbulence that allow us to estimate the size, strength and spatial distribution of the sources as they move downstream.

Experimental studies of turbulence normally consist of a velocity or other time history records obtained at fixed points which are difficult to interpret in terms of unsteady complex patterns of motion. When one examines experimental records, which are obtained from a succession of disturbances passing the measuring stations, information from those parts which are relatively remote from each other, either in space or time, does appear to be statistically independent. Thus the measured signals are normally regarded as random variables and described in statistical terms.

The adoption of a statistical approach represents a substantial experimental simplification, both in the number of observations necessary and in their subsequent analysis. However, though statistical descriptions provide a concise, systematic and convenient method of presenting experimental observation, they tend to be inadequate for the evaluation of source strength characteristics, since they average out many significant features of the motion. For example, statistical descriptions include the mean Reynolds stresses $\rho v_i v_j$ at a point, with perhaps their amplitude spectrum. The direct evaluation of source integrals in (18) or (20) requires a knowledge of their spatial complex rate of change (both amplitude and phase) in a moving volume of fluid associated with a moving source. (see e.g. Section A1.1)

A1.5.1 Correlation length and time scales

One can, however, obtain useful information concerning the length scales of the turbulent patterns and their rates of change from velocity correlation measurements. Given two velocity signals u_i and u_j, with standard derivations σ_i and σ_j, obtained from two points with a separation vector ξ, their cross correlation function can be defined as

$$R_{ij}(\xi,\tau) = \frac{1}{T} \int_{0}^{T} \frac{u_i(x,t) \cdot u_j(x+\xi,t+\tau)}{\sigma_i \sigma_j} \, dt, \qquad \text{A(22)}$$

where τ is the time separation introduced between the two records during analysis. Since velocity patterns are continuous, we expect high correlation between signals with small separations in space or time, while the magnitude of the correlation decays as the separation increases.

Correlations performed with zero time delay describe spatial characteristics, those performed with zero space separation (autocorrelations) describe fixed point temporal characteristics. Space-time correlations, as in (22), may be employed to describe the temporal characteristics of a time averaged moving pattern (reference 3).

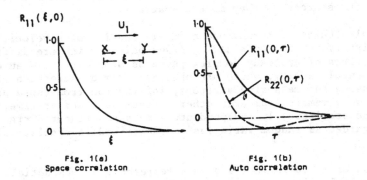

Fig. 1(a)
Space correlation

Fig. 1(b)
Auto correlation

FIG. 1(a)   FIG. 1(b)
Space Correlation   Auto Correlation

Measurements of the type shown can be used to derive mean space and
time scales of the eddy pattern.   Thus the eddy length scale L_ξ is
normally defined as

$$ L_\xi = \int_0^\infty R(\xi,o)\,d\xi \ . \qquad\qquad\qquad A(23) $$

A similar expression can be evaluated to describe a time scale

The interpretation of the auto-correlation function is straightforward.   Since the velocity field is continuous, the auto-correlation
at a point is a function only of the time delay τ, so must be symmetric
in time, or an even function of τ.   For example the auto-correlation in
Fig. 1(b) can be transformed to yield the frequency power spectrum of
the signal, or

$$ G_{11}(o,\omega) = \int_{-\infty}^\infty R_{11}(o,\tau)\exp(-i\omega t)\,d\omega \ , \qquad\qquad A(24) $$

where ω is the radian frequency.   Only if a moving pattern of turbulence
is frozen, thus it generates no sound (when it is moving subsonically),
can a similar interpretation be placed on the space correlation, in terms
of a wave number spectrum.

To find the temporal rate of change of a pattern, in a frame moving
with the average speed of the velocity patterns, one can transform the
envelope of the space time correlations shown in Figure 2.

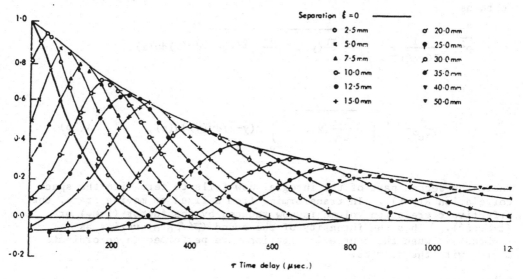

FIG.2   Cross-Correlation Measurements in Jet Mixing Region

Here the separation direction ξ has been chosen so that the pattern
convects, over the separation interval, at a constant speed u_c.   This
is the velocity of propagation of mean turbulent energy and is therefore
analogous to a group velocity, though it is not the phase speed of the
individual spectral components.   We have seen that some of these must
have phase speeds that are supersonic to produce a radiated sound field.
This information provides the frequency – suitably adjusted by necessary
doppler factors – of the radiating source, but not its magnitude.

A1.5.2 Intensity of turbulence generated sound

Lighthill suggested that, in so far as only motions at nearby points
in a turbulent velocity field were well correlated, this fact could be
used to simplify the evaluation of turbulent sources.   Thus each
correlated volume would contribute separately but that one need only
perform the detailed evaluation over each of the correlated volumes.
The time-averaged pressure correlations for an observer at x could be
expressed as

$$\overline{p^2} = \overline{p(x,t)p(x,t+\tau)},$$

where the time bars represent time averages, c.f. (22).

Making use of (20), and ignoring the motion for the moment, one obtains

$$\overline{p^2} = \frac{1}{16\pi c_o^2 r^2} \int\limits_V \int\limits_V \overline{\frac{\partial^2 Trr}{\partial t^2}(\tilde{y},t'), \frac{\partial^2 Trr}{\partial t^2}(\tilde{z},t')} \, dv(y) \, dv(z),$$

$$\approx \frac{1}{16\pi c_o^2 r^2} \int\limits_V \overline{\left(\frac{\partial^2 Trr}{\partial t^2}(\tilde{y},t) \right)^2} \int\limits_V R(\tilde{y}-\tilde{z}) \, dv(y) \, dv(z), \qquad A(25)$$

where the second part of the integral now falls to zero with the space correlation R(y−z). In transforming from a fixed to moving frame, the effective correlation volume is now reduced by the factor $(1-M_r)$ or $(1-M\cos\theta)$. Thus the intensity of the sound $(\overline{p^2}/\rho_o c_o)$ varies as $(1-M\cos\theta)^{-5}$ when the source integrations are performed in co-ordinates moving with the sources.

References

1.   Lighthill, M.J. 1952 On sound generated aerodynamically I General Theory, Proceedings Royal Society A211, pp 564 - 587

     II Turbulence as a source of sound, Proceedings Royal Society A222, pp 1 - 32.

2.   Lighthill, M.J. 1961 The Bakerian Lecture 1961. Sound generated aerodynamically. Proceedings Royal Society A267, 147 - 182.

3.   Davies, P.O.A.L. 1973. Structure of turbulence. Journal Sound and Vibration Vol.28, pp 513 - 526.

Structure-Borne-Sound

M. Heckl, Technische Univer-
sität Berlin

Although sound waves in structures cannot be heard directly,
and only be felt at low frequencies, they play an important
role in noise control, because many sound signals are generated or transmitted in structures before they are radiated
into the surrounding medium. In several respects sound waves
in structures and sound waves in gases or liquids are similar, there are, however, also fundamental differences, which
are due to the fact that solids have a certain shear stiffness, wheras gases or liquids show practically none. As a
consequence acoustic energy can be transported not only
by the normal compressional waves but also by shear waves
and many combinations of compressional (sometimes loosely
called longitudinal) and shear waves . For noise control
purposes bending waves (which are a special combination of
compressional and shear waves) are of primary importance;

for some special cases (quasi-) longitudinal waves and
torsional waves also have to be considered.

1. Basic equations

The governing equations and wave speeds are /1, 2/:

bending waves:

$$B \left(\frac{\partial^2}{\partial x^2} + \frac{\partial^2}{\partial y^2} \right)^2 v + m'' \frac{\partial^2 v}{\partial t^2} = \frac{\partial p}{\partial t}$$

$$w = \frac{\partial v}{\partial x}; \quad \frac{\partial M}{\partial t} = - B \frac{\partial^2 v}{\partial x^2}; \quad F = - \frac{\partial M}{\partial x} \tag{1}$$

$$c_B = \sqrt[4]{\omega^2 \, B/m''}$$

valid if $f < c_{L1}/20 \ h$

(quasi-)longitudinal waves in bars:

$$Y \frac{\partial^2 v}{\partial x^2} - \rho \frac{\partial^2 v}{\partial t^2} = \frac{\partial p'}{\partial t} \tag{2}$$

$$c_{L1} = \sqrt{Y/\rho}$$

valid if $f < c_{L1}/ 6 \ h$

torsional waves in bars:

$$T \frac{\partial^2 w}{\partial x^2} - \theta \frac{\partial^2 w}{\partial t^2} = - \frac{\partial M'}{\partial t} \tag{3}$$

$$c_T = \sqrt{T/\theta}$$

Rayleigh, surface waves:

$$c_R \approx 0,92 \sqrt{G/\rho} \tag{4}$$

valid if $f \gg c_{L1}/20 \ h$

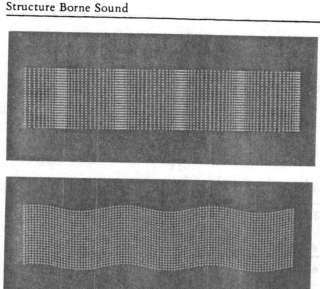

compressional wave
in infinite medium

shear wave in
infinite medium

(quasi-) longitudi-
nale wave in bar

bending wave in
beam or plate

Rayleigh
surface wave

Fig. 1: Some wave types in solids

orthotropic plates:

$$B_x \frac{\partial^4 v}{\partial x^4} + B_{xy} \frac{\partial^4 v}{\partial x^2 \partial y^2} + B \frac{\partial^4 v}{\partial y^4} + m'' \frac{\partial v^2}{\partial t^2} = \frac{\partial p}{\partial t} \tag{5}$$

waves in cylindrical, thin shells

$$m'' \frac{\partial^2 v_r}{\partial t^2} + Y \frac{h^3}{12} \left(\frac{1}{a^2} \frac{\partial^2}{\partial \varphi^2} + \frac{\partial^2}{\partial z^2} \right)^2 v_r + \frac{Yh}{a} \left(\frac{v_r}{a} + \mu \frac{\partial v_z}{\partial z} + \frac{\partial v_\varphi}{a \partial \varphi} \right)$$
$$= \frac{\partial p}{\partial t}$$

$$m'' \frac{\partial^2 v_z}{\partial t^2} - Yh \left(\frac{\mu}{a} \frac{\partial v_r}{\partial z} + \frac{\partial^2 v_z}{\partial z^2} + \frac{1-\mu}{2} \frac{\partial^2 v_z}{a^2 \partial \varphi^2} + \frac{1+\mu}{2} \frac{\partial^2 v_\varphi}{a \partial z \partial \varphi} \right) = 0 \tag{6}$$

$$m'' \frac{\partial^2 v_\varphi}{\partial t^2} - Yh \left(\frac{1}{a^2} \frac{\partial v_r}{\partial \varphi} + \frac{\partial^2 v_\varphi}{a^2 \partial \varphi^2} + \frac{1-\mu}{2} \frac{\partial^2 v_\varphi}{\partial z^2} + \frac{1+\mu}{2} \frac{\partial^2 v_z}{a \partial z \partial \varphi} \right)$$

In these equations: B = bending stiffness, m" = mass per unit
area, c = wave speed (phase speed) Y = Young's modulus,
ρ = density, p = exciting pressure, p' = exciting force per
length, T = torsional stiffness, w = angular velocity,
θ = moment of inertia, M' = exciting moment per length,
G = shear modulus, a = radius of cylinder, v_r, $v\varphi$, v_z = components of shell wall velocity, n = 1,2, 3... μ = Poisson's
ration, M = bending moment, F = force in beam when bending.

2. Bending wave field on beams and plates

For a beam eq. 1 becomes for pure tones

$$B \frac{d^4 \hat{v}}{dx^4} - \omega^2 m' \hat{v} = j\omega \hat{p}. \tag{7}$$

Thus the natural solutions are

$$\hat{v} = v_+ e^{-jk_B x} + v_- e^{+jk_B x} + v_{j+} e^{-k_B x} + v_{j-} e^{k_B x} , \tag{8}$$

where $k_B = \sqrt[4]{\omega^2 m'/B}$. The first two terms are travelling
waves in both directions. The other two terms are exponentially decaying near fields. The factors v_+, v_-, v_{+j}, v_{j-}

have to be adjusted in such a way that the boundary conditions are fulfilled. For a semiinfinite beam driven at
the end by a force F_o the conditions are

$$v_- = v_{j-} = 0; \quad \hat{M}(0) = 0, \quad \hat{F}(0) = \frac{B}{j\omega} \frac{\partial^3 v(0)}{\partial x^3} = \hat{F}_o \qquad (9)$$

giving

$$\hat{v} = \hat{F}_o \frac{\omega}{Bk_B^3 (1+j)} \left(e^{-jk_B x} + e^{-k_B x} \right). \qquad (10)$$

Eq. (10) can be used to find the input impedance which is
defined by

$$Z = \frac{\hat{F}_o}{\hat{v}(x = 0)} \qquad (11)$$

Thus for a semi-infinite beam the relation

$$Z = \frac{Bk_B^3}{2\omega} (1+j) = m'c_B \frac{1+j}{2} \qquad (12)$$

holds.

For an infinite beam driven in the "center" the impedance
is

$$Z = 2 m'c_B (1 + j) \qquad (13)$$

For a plate the situation is somewhat more complicated.
It can, however, be shown /1/ that the velocity field of
an infinite plate, which is excited by a point force $\hat{F}_o$
is given by

$$\hat{v}(r) = \frac{\hat{F}_o}{8\sqrt{Bm''}} \left[H_o^{(2)}(k_B r) - H_o^{(2)}(-j k_B r) \right]. \qquad (14)$$

Here r is the distance between the measuring point and
the point of excitation, $H_o^{(2)}(\ldots)$ is the Hankel function

of the second type and zero order. Since

$$H_o^{(2)}(k_Br) - H_o^{(2)}(-jk_Br) \approx 1 \text{ for } k_Br \ll 1$$

the input impedance of a plate is

$$Z = 8\sqrt{B\ m''} \quad . \tag{15}$$

Therefore the mechanical power P transmitted by a point

force into a plate is

$$P = \frac{1}{2} \text{Re}\left\{\hat{F}_o \cdot \hat{v}(0)\right\} = \frac{1}{2}\left|\hat{F}_o^2\right| \text{Re}\left\{\frac{1}{Z}\right\} = \frac{\left|\hat{F}_o^2\right|}{16\sqrt{B\ m''}} \quad . \tag{16}$$

If an infinite plate is excited by any pressure distribu-

tion, the principle of superposition can be applied; i.e.

the pressure $\hat{p}(x,y)$ is decomposed into many little forces

of amplitude $\hat{p}(x_q, y_q)\ dx_q\ dy_q$ and the velocity fields

generated by each of these little forces is added. This

way we get

$$\hat{v}(x,y) = \frac{1}{8\sqrt{B\ m''}} \int \hat{p}(x_q, y_q)\left[H_o^{(2)}(k_B\ r_q) - H_o^{(2)}(-jk_Br_q)\right]dx_q dy_q$$

where

$$r_q = \sqrt{(x - x_q)^2 + (y - y_q)^2} \quad .$$

In some distances from the excitation the asymptotic expan-

sion of the Hankel function can be used giving:

$$v(x,y) = \frac{1}{8\sqrt{B\ m''}} \sqrt{\frac{2}{\pi k_B}}\ e^{\ j\ \pi/4} \int \hat{p}(x,y)\frac{1}{\sqrt{r_q}}\ e^{-jk_Br_q}\ dx_q\ dy_q \tag{18}$$

3. Energy considerations

Quite often general conclusions ca be obtained by energy

considerations in structures. This is especially true when
broad frequency bands are excited and when several resonances lie in the frequency band of interest. In such cases
the energy transfer is determined by the impedance of the
corresponding infinite system, thus all the complications
associated with finding eigenfunctions and eigenfrequencis
can be avoided. The disadvantage is that the results obtained this way hold only in an average sense and may be rather
inaccurate when narrow frequency bands are exited.

The structure borne sound energy E that is generated by an
impulsive source (hammer blow, punching machine, piston
slap, etc.) is given by

$$E = \int F(t)\, v(t)\, dt \ . \tag{19}$$

Here F(t) ist the time history of the exciting force and
v(t) the velocity at the point of excitation. Taking the
Fouriertransforms of F, v and E, gives the following relation for the energy E(ω) at the frequency ω:

$$\check{E}(\omega) = \check{F}(\omega) \cdot \check{v}(-\omega) = \left| \check{F}(\omega) \right|^2 . A(\omega) . \tag{20}$$

Here

$$\check{F}(\omega) = \int F(t)\, e^{-j\omega t}\, dt; \quad \check{v}(\omega) = \int v(t)\, e^{-j\omega t}\, dt . \tag{21}$$

The quantity A(ω) is the real part of the input admittance
at the point of excitation. It is related to the input
impedance defined in eq. (12) by

$$A(\omega) = \mathrm{Re}\left\{ \frac{\hat{v}}{\hat{F}_0} \right\} = \mathrm{Re}\left\{ \frac{1}{Z} \right\} \ . \tag{22}$$

For the two important cases (eq. 12, 13 and 15) that have
been considered already and for many other examples A(ω)

Another method to reduce the generation of structure borne
energy is to decrease the input admittance; i.e. to increase
the impedance (eq. 22). This can be achieved by using heavy,
stiff and highly damped structures where the force is
applied. (Fig. 3)

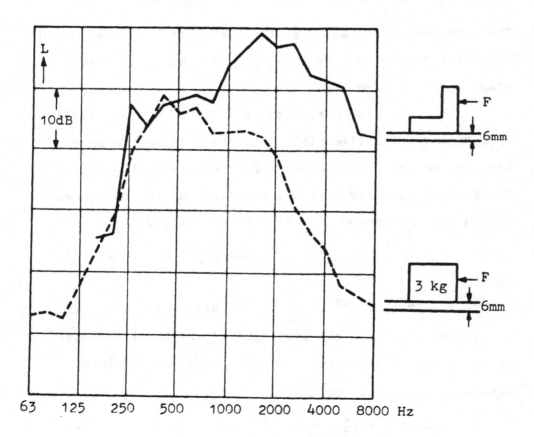

Fig. 3: Influence of an added mass (increasing the input
        impedance) on the radiation from a 6 mm steel plate
        Excitation by five impacts per second.

4. Energy considerations for coupled structures

Let us assume a structure consisting of two parts that
are connected in some way.

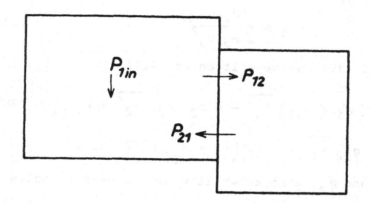

Fig. 4: Coupled structure

If the structure is separated from its surrounding and if
steady state conditions hold, the law of conservation of
energy gives the following balance equations

$$P_{1e} + P_{12} = P_{1\ in} + P_{21} \qquad (24)$$
$$P_{2e} + P_{21} = P_{12} \quad .$$

Here P_{1e}, P_{2e} are the mechanical powers converted to heat
in the two parts, $P_{1\ in}$ is the power supplied from the outside, P_{21} and P_{12} are the powers flowing from part 1 to 2
and vice versa.

It will been shown in the next chapter (eq.(42)) that

$$P_{1e} = \omega\ \eta_1\ m_1\ \overline{v_1^2}\ ; P_{2e} = \omega\ \eta_2\ m_2\ \overline{v_2^2}\ ; \qquad (25)$$

where η_1, η_2 are the loss factors, m_1, m_2 the masses and
$\overline{v_1^2}$, $\overline{v_2^2}$ the mean square velocities of part 1 and 2.
Since the power P_{12} can be considered as being lost for

part 1, we can define a "coupling loss factor η_{12}" by

$$\eta_{12} = \frac{P_{12}}{\omega\, m_1\, v_1^{\,2}} \qquad , \qquad (26)$$

and similary

$$\eta_{21} = \frac{P_{21}}{\omega\, m_2\, v_2^{\,2}} \qquad . \qquad (27)$$

Thus eq. (39) can be written as /4,1/

$$\omega\, m_1 \left(\eta_1 + \eta_{12} \right) \overline{v_1^{\,2}} - \omega\, m_2\; \eta_{21}\, \overline{v_2^{\,2}} = P_1 \text{ in} \qquad (28)$$

$$- \omega\, m_1\; \eta_{12}\, \overline{v_1^{\,2}} + \omega\, m_2 \left(\eta_2 + \eta_{21} \right) \overline{v_2^{\,2}} = 0 \; .$$

If η_{12} and η_{21} were known it would be easy to solve eq.(28).
Unfortunately this is usually not the case, but in spite
of that eq. (28) can be very useful for the special case
of strong coupling. The conditions for strong coupling are
fulfilled when $\eta_2 < \eta_{21}$ i.e. when the loss factors for
energy transfer into heat are low - as is usually the case
for metal constructions.

Under these conditions we find

$$\frac{\overline{v_2^{\,2}}}{v_2^{\,2}} \approx \frac{m_2}{m_1}\; \frac{\eta_{21}}{\eta_{12}} \qquad . \qquad (29)$$

By applying the principle of reciprocity it can be shown
/1/, that η_{21}/η_{12} depends only on the number of modes
(resonances) ΔN_1 and ΔN_2 that lie in the excited frequency band $\Delta\omega$. Therefore we find /4/:

$$\frac{\overline{v_1^{\,2}}}{v_2^{\,2}} \approx \frac{m_2}{m_1}\; \frac{\Delta N_1}{\Delta N_2} \quad (\text{for } \eta_2 < \eta_{21}) . \qquad (30)$$

Examples for ΔN are:

Bending waves on beams: $\dfrac{\Delta N}{\Delta \omega} = \dfrac{1}{2\pi} \sqrt{\dfrac{\varrho}{K^2 \omega^2 Y}}$

Bending waves on plates: $\dfrac{\Delta N}{\Delta \omega} = \dfrac{S}{2\pi h} \sqrt{\dfrac{3}{Y} \varrho}$.   (31)

Here ρ is density of the material, Y is Young's modulus,
h the plate thickness, S the plate area, l the length of
the beam and K its radius of gyration.

As an interesting consequence one can see from eq. (30) that
the velocity v_2 - i.e. of that part that is not excited
directly - can be larger than v_1. (Fig. 5) This is expeci-
ally true when part 2 consists of thin, undamped sheet
metal. The deteriorating effect of a strongly coupled se-
condary part of a construction is quite often encountered
in practice when a solid piece of machinery is rigidly
connected to a light, undamped covering.

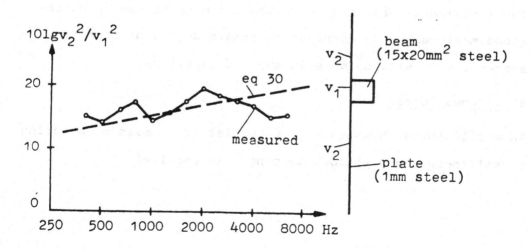

Fig. 5: Velocity ratio of a strongly coupled beam-plate
        system. The excitation was on the beam.

5. General laws of vibration damping

The safest way to get rid of acoustic energy is to transform
it into heat. For structure borne sound this can be done most
effectively by applying damping materials such as special
high polymers or grainy materials such as sand or by making
use of friction losses at junctions, interfaces, etc.. Obviously damping materials are most effective when they are
applied in the immediate vicinity of the source because the
amount of energy transformed into heat is especially large
at those places where the vibration amplitudes are highest.

The basic mechanisms that give rise to damping are dry friction, viscous losses, and relaxation mechanisms. Whereas
the details of vibration damping by dry friction is not yet
fully understood, there is a fair amount of knowledge on
damping by viscosity and relaxation.

The fundamental processes of vibration damping can be understood most easily by considering single degree of freedom
systems. This will be done in the next sections.

a) Viscous losses

An oscillator as hown in fig. 6 consists of a mass m, a spring
of stiffness s and viscous damping r in parallel.

does not change very much with frequency, therefore we can assume

$$\check{E}(\omega) \sim \left| F(\omega) \right|^2$$

and using Parseval's relation we find

$$E \sim \int F(t)^2 \, dt \qquad . \qquad (23)$$

Eq. (23) shows that for a given momentum $\int F(t)dt$ of an impact the energy generation decreases with increasing duration of the impact. Therefore everything that makes the duration of an impact longer helps to reduce the noise; this is especially true for the higher frequencies. Thus we find the welknown rule, that sudden changes cause annoying high frequency noise, whereas continuous changes generate mainly low frequencies.

In practical situations continous changes can be achieved by using toothed wheels with diagonal cuts, inclined working tools (see fig. 2), smooth transitions at joints, etc., etc. or by applying resilient interlayers.

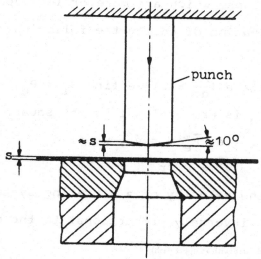

Fig. 2: Punch press with inclined tool

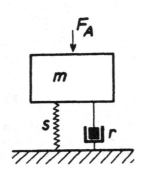

If F_A is the exciting force, the displacement ξ of the mass is given by

$$m \ddot{\xi} + r \dot{\xi} + s \xi = F_A \quad (32)$$

The force F transmitted into the foundation is

$$F = s \xi + r \dot{\xi} . \quad (33)$$

Fig. 6: Mechanical oscillator

Multiplying eq. (32) by $\dot{\xi}$ we get the power balance equation

$$\frac{d}{dt} (\frac{1}{2} m \dot{\xi}^2 + \frac{1}{2} s \xi^2) = F_A \dot{\xi} - r \dot{\xi}^2 , \quad (34)$$

or in a more general way

$$\frac{d}{dt} (E_{kin} + E_{pot}) = P_i - P_e . \quad (35)$$

Here E_{kin} is the kinetic energy, E_{pot} the potential energy, P_i the power supplied by the source, P_e is the power that is lost, i.e. transformed into heat and t is the time.

From the above equations, which are more or less applications of the law of conservation of energy the following can be deduced: [+]

α) In the steady state ($\frac{d}{dt} = 0$) we find $\bar{P}_i = \bar{P}_e$.

Since usually P_e is proportional to the square of the velocity $v = \dot{\xi}$; i.e. $\bar{P}_e = r \bar{v}^2$, we find

$$\bar{v}^2 = \bar{P}_i / r . \quad (36)$$

The square of the velocity of a vibrating system in the steady state is inversely proportional to the damping.

[+] (With reference to r.m.s. values, indicated with an upper dash).

β) If the vibrating system is suddenly disconnected from the source ($P_i = 0$), the motion dies down exponentially. Since for free vibrations and small damping $\bar{E}_{kin} \cong \bar{E}_{pot}$ we find

$$\frac{d}{dt} m\, \bar{v}^2 \cong - r\, \bar{v}^2; \text{ or } \bar{v}^2 \cong \bar{v}_o^2\, e^{-\frac{r}{m} t} \qquad (37)$$

γ) For a purely periodic motion (e.g. $\xi = \xi_o \cos \omega t$) there is a phase shift of $\tan^{-1} (\omega r/s)$ between force and displacement. See eq. (33).

δ) As is well known damping "brings down" and rounds off the resonance peaks.

ε) For the particular case of viscous damping the total force transmitted into the foundation at very high frequencies increases with damping, because the impedance r of the damper eventually becomes larger than the impedance s/ω of the spring.

b) Relaxation

The stress-strain relation underlying eq. (33) does not give a good description of the behaviour of springs made out of rubber or similar materials. In this case it is better to assume, that the force acting on the spring depends not only on the actual displacement but also on the displacement at earlier times. Thus the material is assumed to have a "memory". Mathematically this can be written as

$$F(t) = s' \cdot \xi(t) - \int_0^\infty \xi(t-\tau)\, \varphi(\tau)\, d\tau \qquad (38)$$

(instead of eq. (33). The simplest case of a "memory function" is

$$\varphi(\tau) = \frac{s_1}{\tau_r} e^{-\tau/\tau_r} , \qquad (39)$$

where τ_r is the relaxation time.

Using these relations one finds for the equation of motion
for a mass resting on a spring with relaxation:

$$m \, \ddot{\xi}(t) + s' . \xi(t) - \int_0^\infty \xi(t - \tau) \, \varphi(\tau) \, d\tau = F_A(t). \qquad (40)$$

A complete solution of this equation is somewhat difficult,
but for purely periodic motion results can be given. It can
be shown, for example, that for materials that are normally
used the stiffness increases somewhat with frequency and
that the damping has a maximum at $\omega = 1/\tau_r$. The energy
losses, the decay rate and the resonance effects are the
same as discussed in the previous section, provided one
replaces r by $s_1 \cdot \tau_r / (1 + \omega^2 \tau_r^2)$. The only exception is
that the total force transmitted into the foundation at
high frequencies is not increased when the losses are increased.

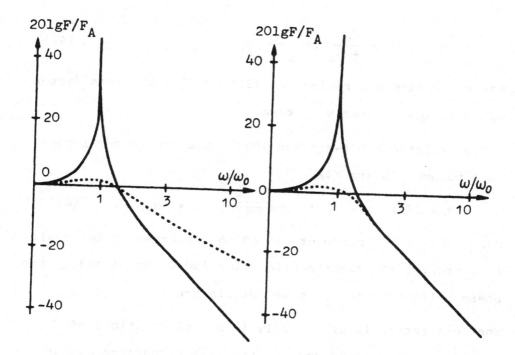

Fig. 7: Force transmission for a mechanical oscillator.

Left: spring with viscous damping in parallel.

Right: spring with relaxation

small losses; ... high losses $\omega_o = \sqrt{s/m}$.

c) Loss factor

Usually the damping occuring in lossy materials is characterized by an energy ratio. This quantity is called the
loss factor and is defined by

$$\eta = \frac{E_e}{2 \pi E_R} \quad . \qquad (41)$$

Here E_e is the energy lost per cycle of duration T, E_R is
the reversible energy. Since $E_e = F_e T$ and $\omega = 2\pi/T$, eq. (41)
can also be written as

$$\eta \; = \; \frac{P_e}{\omega\, E_R} \; \approx \; \frac{P_e}{\omega\, m\, \overline{v}^2} \qquad . \tag{42}$$

For the system underlying eq. (32) or (33) the loss factor
would be $\eta = r\, v^2/\omega m\, v^2 = r/\omega m$.

Useful relations between the loss factor and other important parameters are /1/

$$\eta = \Delta f/f; \quad \eta = 1/Q; \quad \eta = \frac{2.2}{Tf}; \quad \eta = tg\,\chi \qquad . \tag{43}$$

Here Δf is the bandwidth (3 dB down points), Q the quali-
ty factor, T the reverberation time (60 dB down) and χ the
phase shift between force and displacement.

The loss factor is used widely in all calculations of
structure borne sound because the loss mechanisms can be
taken into account in all equations of motion, as long as
it is periodic, simply by multiplying the elastic modul
by $(1 + j\eta)$. Thus instead of the stiffness s a complex
stiffness $s(1 + j\eta)$ and instead of Youngs modulus Y the
quantity $Y(1 + j\eta)$ are introduced.

For some materials the loss factors and other important
quantities are given in the following table

Material	Density	Y	μ	c_L	η
Aluminium	2700	72.10^9	0,34	5200	$\approx 10^{-4}$
Lead	11300	17.10^9	0,43	1250	$\approx 2.10^{-2}$
Cast Iron	7800	200.10^9	0,3	5000	$\approx 10^{-3}$
Steel	7800	200.10^9	0,3	5100	$\approx 10^{-4}$
Copper	8900	125.10^9	0,35	3700	$\approx 10^{-3}$
Brass	8500	100.10^9	0,33	3200	$< 10^{-3}$
Glass	2500	60.10^9		4900	$\approx 10^{-3}$
Wood	400-800	$1-5.10^9$		2000-3000	$\approx 10^{-2}$
Concrete	2300	20.10^9		3500	$\approx 5.10^{-3}$

The density is given in kg/m^3, Young's modulus in N/m^2 (10^9 N/m^2 = 10^4 kp/cm^2). μ is Poisson's ratio and c_L the speed of longitudinal waves in m/s.

For high polymers data cannot be given, because in the important so called transition region they depend very much on temperature and other parameters. It should be noted however, that good materials used for damping layers usually have a Young's modulus of app. 10^9 N/m^2 and a loss factor around η = 1. High polymers used in sandwich plates are softer and have loss factors higher than unity.

The loss factors given in the table hold for specimens consisting of one piece and having no energy transfer into the neighbourhood. Real life structures made out of metals usually have higher losses than the materials alone,

because there is always some friction at interfaces etc..
Typical values for constructions without special damping
treatment are /5/:

   a) constructions made out of a few fairly thick pieces
      (e.g. ship hull)

$$\eta \approx 3 \cdot 10^{-3} \quad \text{for} \quad f < 500 \text{ Hz}$$
$$\eta \approx 10^{-3} \quad \text{for} \quad f > 1000 \text{ Hz},$$

   b) constructions made out of a few thin, or many thick
      pieces (e.g. car body, or diesel engine)

$$\eta \approx 10^{-2},$$

   c) constructions made out of many thin pieces

$$\eta \approx 5 \cdot 10^{-2} \quad \text{for} \quad f < 500 \text{ Hz}$$
$$\eta \approx 10^{-2} \quad \text{for} \quad f > 1000 \text{ Hz} \quad .$$

Obiously any additional damping treatment is useful only
when it increases the loss factor above these values.

6. Damping layers

Damping layers are high polymers that are sprayed, painted
or otherwise spread into metal constructions, especially
to plates in bending, in order to reduce the vibration
amplitudes. As can be seen from the figure the layer is
strechted and compressed as the plate undergoes bending
motion. /8, 1/

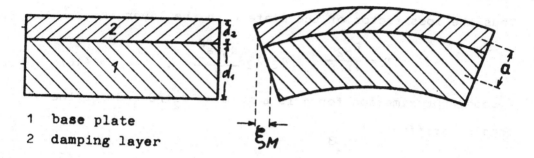

1 base plate
2 damping layer

Fig. 8: Bending of plate with damping layer

Therefore the energy transferred into heat in the damping layer is (see eq. 41)

$$E_{e2} = 2 \pi \eta_2 E_{R2} = \pi \eta_2 Y_2 d_2 \left| \frac{d \xi_m}{dx} \right|^2 \quad . \qquad (44)$$

Y_2 is Young's modulus, η_2 the loss factor, d_2 the thickness of the damping layer, $d \xi_m/dx$ is the amount by which the center line of the layer is strechted or compressed. The above equation shows quite clearly that in order to achieve large losses, $Y_2 \cdot \eta_2$ has to be big, the damping layer should be thick and the motion of its central line should be large. (There must be not slip at the interface between plate and layer.)

Since there are practically no losses in the base plate, the loss factor of the combination is obtained by comparing E_{e2} with the reversible energy of the whole system which is

$$E_R = \frac{1}{2} B \left| \frac{d \beta}{dx} \right|^2 \quad , \qquad (45)$$

where B is the bending stiffness of the combination and β the angle of flexure. If a is the distance from the neutral fiber to the center line of the damping layer then $\xi_m = a\beta$. Thus the loss factor of the plate with the layer is

$$\eta = \frac{E_{e2}}{2\pi\,E_R} = \eta_2\,Y_2\,\frac{d_2\,a^2}{B} \quad . \tag{46}$$

A good approximation for a is $a \approx (d_1 + d_2)/2$ and for the bending stiffness

$$B \approx Y_1\,\frac{d_1^3}{12} + Y_2\,d_2\,a^2 \quad . \tag{47}$$

The values for η can also be taken from fig. 9. It can be seen, that η increases approximately with the square of d_2; thus the damping layer should be rather thick and never be divided to both sides of the plates. Typical values for a good damping layer on a steel plate are $Y_2 \cdot \eta_2 \approx 10^9$ N/m^2; $d_2 = 2\,d_1$; $\eta \approx 0,1$. Applied to stiff beams (e.g. I-beams) the loss factor is smaller because B appears in the denominator of eq. (46).

It should be noted, that for longitudinal waves the loss factor is much smaller, than the values given by eq. (46). The vibration reduction that can by obtained by damping layers depends on the type of motion and on the loss factor η_B that existed already before the damping was applied.

If a system is undergoing high frequency stick-slip vibrations (e.g. break squeal), the damping may interrupt the stick-slip feedback mechanism and reduce the vibrations

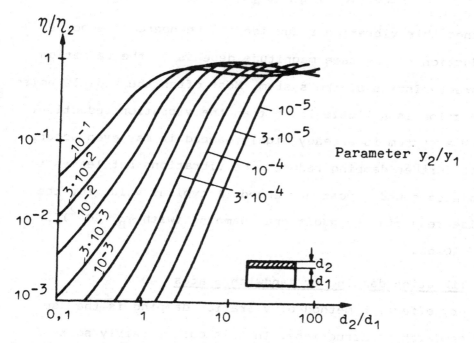

Fig. 9: Loss factor of plates with damping layers

by 20 or 30 dB. Unfortunately it cannot be predicted wether
this effect occurs or not.

If a system is excited at its resonance frequency, additional damping reduces the amplitude by a factor η / η_B or

$$\Delta L = 20 \lg \eta / \eta_B \quad dB, \qquad (48)$$

where η_B is the damping before and η after application of
the treatment.

If — and this is the most common case — a system is excited
by random forces, additional damping reduces the resonance
amplitude, but it also makes the resonance curve broader,
therefore the vibration reduction is

$$\Delta L \approx 10 \lg \eta / \eta_B \quad \cdot \qquad\qquad (49)$$

Wether this vibration reduction can be heard as a noise reduction of the same magnitude depends on the radiation characteristics of the system. Most often the audible noise reduction is a little lower than the vibration reduction. When a system is already highly damped it may even occur, that further damping reduces the vibrations but leaves the radiated sound almost unchanged. A typical value for the noise reduction by additional damping (with $\eta_B \simeq 10^{-2}$) is 7 - 10 dB.

7. Vibration damping by sandwich plates

A very effective method of vibration damping is the use of sandwich constructions. In this case a fairly soft material is placed between two stiff plates. As the plates bend, the intermediate layer undergoes a shearing motion and if its loss factor is high a certain amount of vibratory energy is transformed into heat.

A complete derivation of the equations of motion would be too lengthy here, therefore only the main results of the theory are stated /1, 6, 7/. With the symbols shown in Fig. 10 it can be shown that

$$\eta = \frac{\eta_2 \cdot X \cdot X_1}{1 + (2 + X_1) X + (1 + X_1)(1 + \eta_2{}^2) X^2} \quad , (50)$$

where

$$\frac{1}{X_1} = \frac{y_1 d_1{}^3 + y_3 d_3{}^3}{12 d_{13}{}^2} \left(\frac{1}{y_1 d_1} + \frac{1}{y_3 d_3} \right) , \qquad (50a)$$

$$X = \frac{G_2}{k^2 d_2} \left(\frac{1}{Y_1 \, d_1} + \frac{1}{Y_3 \, d_3} \right) \quad , \qquad (50b)$$

$$k^2 = \omega \sqrt{\frac{12(\rho_1 \, d_1 + \rho_3 \, d_3)}{(Y_1 \, d_1^{\,3} + Y_3 \, d_3^{\,3}) \, (1 + X \, X_1 / (1 + X))}} \quad . \quad (50c)$$

Here ρ_1, ρ_3 are the densities und G_2 the shear modulus of the damping material.

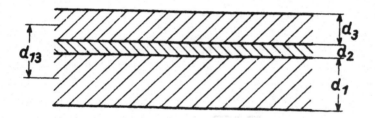

Fig. 10: Sandwich plate

The equations show that the loss factor depends on frequency and has a very broad maximum at

$$f_{max} \approx \frac{1}{2\pi} \frac{G_2}{Y_3 \, d_3 \, d_2} \left(1 + \eta_2^{\,2} \right) \sqrt{\frac{Y_1 \, d_1^{\,2}}{12 \, \rho_1}} \quad . \qquad (51)$$

The loss factors at this frequency can go up to 0,7.
Examples for the noise reduction achieved by the use of sandwich plates (in rather ideal cases) are:

railway bridge made out of thin steel plates ΔL = 13 dB(A)

packing machine for nuts and bolts ΔL = 14 dB(A)

conveyer (shaking trough for metal pieces) ΔL = 10 dB(A)

Since the loss factor and the shear modules of damping

materials change considerably with temperature, great care

has to be taken that a material with a wide temperature

range is used and that the specified temperature range is

not exceeded.

8. Damping by localised absorbers

For several applications vibration damping is achieved by

resonators, that are attached to the vibrating body and

therefore draw out vibratory energy and partly convert it

into heat.

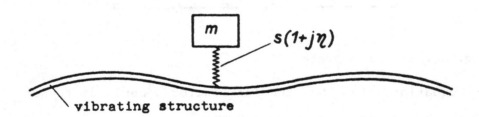

Fig. 11: Plate with resonator

The mechanism of such a system can be understood quite

easily by considering the mechanical impedances of the

original system and the resonator.

The impedance of the resonator is

$$z_1 = j\omega m \frac{1}{1 - \omega^2 \, m/s \, (1 + j \eta)} \qquad . \qquad (52)$$

The power that is lost in the resonator is

$$P_e = \frac{1}{2} Re \left\{ F_1 \cdot v_1 \right\} = \frac{1}{2} \left| v_1 \right|^2 Re \left\{ Z_1 \right\} . \qquad (53)$$

It also has to be taken into account that the velocity v_1 at the point of attachment is somewhat different from the original velocity v_0, because the resonator is a load for the original system therefore

$$\frac{v_0}{v_1} = \frac{Z_0 + Z_1}{Z_0} , \qquad (54)$$

where Z_0 is the point impedance of the original system. Combining the equations yields

$$P_e = \frac{1}{2} v_0^2 \left| \frac{Z_0}{Z_0 + Z_1} \right|^2 Re \left\{ Z_1 \right\} , \qquad (55)$$

It can be seen that this equation contains two counteracting effects. If $Re \left\{ Z_1 \right\}$ is small, i.e. if the resonator is only lightly damped, P_e is small, too. If Z_1 is very large the quantity P_e is small because $Z_1 \gg Z_0$; there is an impedance mismatch, which reduces v_1 and therefore the losses, too. The highest losses are obtained when $Re \left\{ Z_1 \right\} \approx Z_0$.

The maximum damping of such localized absorbers can be astonishingly high especially at low frequencies. It can be shown that a well adjusted resonator absorbs 50 % of the energy that is contained in a region of $\lambda/6$ radius. Obviously it is not an easy matter to tune a resonator in such a way that the optimum is obtained. Possible applications are at railway-car wheels, or at gear boxes i.e. cases where the excitation is limited to very narrow frequency bands.

9. Attenuation of structure borne sound by discontinuities

9.1 Elastic interlayers

a) Mass spring systems

The most common method of vibration isolation is the elastic

mounting. In its simplest form it can be represented by a

mass-spring system resting on a foundation.

It is known, that such a system has a resonance frequency

given by

$$f_R = \frac{1}{2\pi} \sqrt{\frac{s}{m}} \quad , \qquad (56)$$

(see also eq. 32) and that for $f > \sqrt{2} \cdot f_R$ the force trans-

mitted into the foundation is smaller than the exciting

force. In reality things are somewhat more complicated be-

cause the mass may vibrate in different directions and may

also make a rocking motion. Thus there are six degrees of

freedom and consequently six resonance frequencies that

depend on the mass, the moments of inertia and the posi-

tioning of the springs. Fortunately it is usually suffi-

cient to consider only the "most important" type of motion.

If the spring shows nonlinear behaviour the stiffness s

in eq. (56) has to be replaced by the quantity dF/dx; i.e.

by the slope of the deflection curve. Since almost all

materials get stiffer when the static load is increased it
may well be that the addition of a mass does not decrease
the resonance frequency. Thus it is not surprising that
there exist lower limits for the resonance frequencies
$f_{R\ min}$. Examples are:

steel springs	$f_{R\ min} < 0,2$ Hz
rubber in shear	$f_{R\ min} < 2$ Hz
rubber in compression	$f_{R\ min} \approx 5$ Hz
cork	$f_{R\ min} \approx 20$ Hz
high polymer foam	$f_{R\ min} \approx 20$ Hz
fibre glass	$f_{R\ min} \approx 20$ Hz

Sometimes elastic mountings are somewhat more complicated
consisting of several masses and springs. Examples are
sketched in Fig. 12.

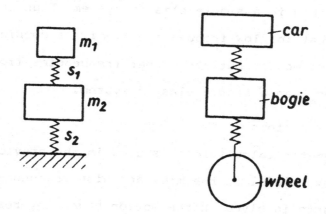

Fig. 12: Examples of double elastic systems.
        left: model of double mount
        right: idealized model of railway car.

Double elastic systems even in the simplest case of motion

in only one direction have two resonances which for the

model shown on the left of Fig. 12 are given by

$$f^2_{I,II} = \frac{1}{2}\left[(f_1^2 + f_2^2 + f_3^2) \pm \sqrt{(f_1^2 + f_2^2 + f_3^2)^2 - 4\,f_1^2\,f_3^2}\right], \quad (57)$$

where

$$f_1 = \frac{1}{2\pi}\sqrt{\frac{s_1}{m_1}}; \quad f_2 = \frac{1}{2\pi}\sqrt{\frac{s_2}{m_2}}; \quad f_3 = \frac{1}{2\pi}\sqrt{\frac{s_1}{m_2}}.$$

The existance of two resonance frequencies may be considered

as a certain disadvantage because the probability of exci-

ting a resonance is increased. Furthermore the problem of

spring stiffening becomes more severe, because at least the

lower spring has to carry the static load of both masses

but the frequency $f_2 < f_I$ is determined by m_2 alone.

But a double mount has also some very useful properties.

It can be shown, that for frequencies above f_I the force

that is transmitted through the system is reduced conside-

rably more than in a single elastic system. Thus one might

summarize that for low frequencies $(f < f_I)$ a double elastic

system may be worse, but for higher frequencies $(f > \sqrt{2} \cdot f_I)$

it is better than a single elastic system.

b) Continous systems

Lumped parameter calculations as used in the previos section

are not very helpful in the mid- and high-frequency range,

when all parts involved in the motion have many resonances.

In this case it is better to apply the concept of mechani-

cal impedance and to treat springs as waveguides. Thus an

indealized system looks like Fig. 13:

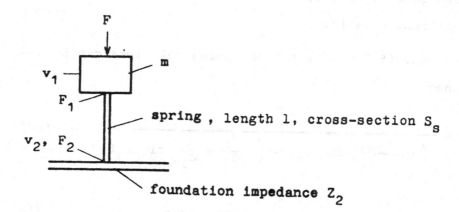

Fig. 13: Elastic mount as a continous system

From Fig. 13 the following equations for pure tone motion (the factor $e^{j\omega t}$ being omitted) can be obtained /1/:

a) for the mass: $F - F_1 = j\omega m\, v_1$ (58)

b) for the spring:

$$F_2 = F_1 \cos k_s l - j\, Z_s\, v_1 \sin k_s l \quad (59)$$

$$v_2 = -\frac{j}{Z_s}\, F_1 \sin k_s l + v_1 \cos k_s l$$

c) for the foundation:

$$F_2 = Z_2\, v_2 \qquad . \qquad\qquad (60)$$

Here v_1, v_2 are the velocities, F, F_1, F_2 the forces, l is the length of the spring, which is assumed to make only longitudinal motion. k_s is the wave number and Z_s the wave impedance of the spring, they are given by

$$k_s = \omega \sqrt{\rho/Y} \qquad\qquad Z_s = S_s \sqrt{Y\rho} \quad , \qquad (61)$$

where Y is Young's modulus, ρ the density and S_s the cross-section of the spring. If the spring moves in bending or torsion or both (e.g. a helical spring) eq. (59)

would look different, but the main conclusions (esp. eq.
(65)) remain valid.

From eqs. (58) - (60) the following relations can be obtained

$$\frac{F_2}{F} = \frac{1}{\left(1 + \frac{j\omega m}{Z_2}\right)\cos k_s l + \left(j\frac{Z_s}{Z_2} - \frac{\omega m}{Z_s}\right)\sin k_s l} \quad , \quad (62)$$

$$\frac{v_2}{v_1} = \frac{1}{\cos k_s l + j\frac{Z_2}{Z_s}\sin k_s l} \quad . \quad (63)$$

If converted to decibels eq. (62) gives the so-called insertion loss, i.e. the reduction of force on the foundation
by adding the mass and the spring. Eq. (63) converted to
decibels gives the difference of the velocity levels on
both sides of the spring. The ratio F_2/F usually is of
prime interest but the quantity that is measured and
plotted easily and therefore most often is v_2/v_1. It is
important to note that $F_2/F \neq v_2/v_1$.

Eq. (62) - (63) depend strongly on the frequency because
of the trigonometric functions appearing in the denominator and obviously there is the danger that the denominator
becomes comparatively small for special frequencies; this
would indicate a rather strong transmission of structure
borne sound into the foundation. Since usually $Z_2 \gg Z_s$
this occurs in the neighborhood of the frequencies given
by $k_s l \approx n\pi$. Using eq. (61) this yields

$$k_s l = \omega_n \, l \sqrt{\frac{\rho}{Y}} = \omega_n \sqrt{\frac{\rho \, S_s \, l}{Y S_s / l}} = \omega_n \sqrt{\frac{m_s}{S}} \approx n\pi \quad . \qquad (64)$$

Here m_s is the mass of the (working part of the) spring. Expressing the stiffness of the spring by eq. (56) we find

$$\frac{\omega_n}{2\pi} = f_n \approx f_R \sqrt{\frac{m}{m_s}} \cdot n\pi \qquad . \qquad (65)$$

Thus the higher order resonance frequencies f_n (where there is poor isolation for structure borne sound) are simply related to a mass ratio. Since very seldom $m > 100 \, m_s$, the first higher order resonance is at most a factor 30 above the frequency f_R. Since the deteriorating effect of the higher order resonances is the more pronounced the smaller the internal damping of the spring it is felt most strongly for metal springs (see Fig. 14). For the first higher order resonance eq. (65) with a slightly different factor holds for more complicated springs as well. Obviously the considerations concerning the transmission of structure borne sound are applicable not only to those springs on which the weight of the engine etc. rests but also for the necessary elastic interlayers at pipes, shafts etc..

Typical improvements (insertion loss) that are achieved in practic by using elastic mounts are 10 - 15 dB, even though the velocity level differences on both sides of the elastic mount may be much greater.

Equation (62) and (63) contain three parameters: the mass m of the vibrating body, the properties of the

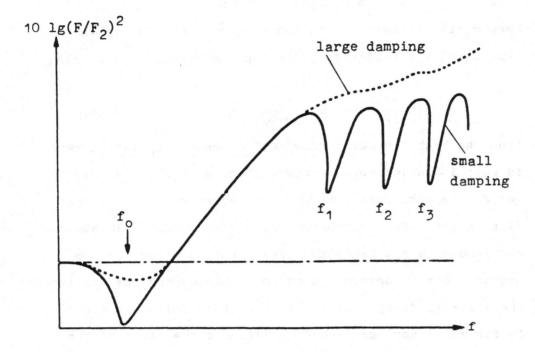

Fig. 14: Influence of higher order resonances on the
         transmission of structure borne sound through
         elastic mounts.

spring and the impedance Z_2 of the foundation. Especially

for fairly soft foundations (e.g. in vehicles or ships)

the influence of the foundation impedance can be of extreme

importance. As a general rule one can say, that an elastic

interlayer is effective only if it is softer, than that

part (usually this part has the dimension of half a ben-

ding wave length) of the foundation that moves in phase

with the excitation point.

9.2 Changes in cross-section, corners, junctions etc.

There are many cases where elastic interlayers cannot be used
for the isolation of structure borne sound because the structural strength is reduced too much. In such cases it is ho-
ped, that other types of discontinuities are helpful. Possible ways for reducing the transmission of structure borne
sound without the use of soft layers are:

damping (see previous chapters)

change of cross-section

change of material

corners or junctions

added masses

In the following table the transmission coefficients for
bending waves on beams are given for several kinds of discontinuities. The transmission coefficient is (for very long
beams) related to the velocities on both sides of the discontinuity by

$$\tau = \frac{P_i}{P_t} = \frac{v_1^2}{v_2^2} \frac{\rho_1 S_1}{\rho_2 S_2} \frac{k_2}{k_1} \qquad . \qquad (66)$$

For plates: $\sigma = h_1/h_2$; for beams: $\sigma = S_1/S_2$. ρ_1, ρ_2 = density,
Y_1, Y_2 = Young's modulus. S_1, S_2 = cross-section,
k_1, k_2 = radius of gyration of the beam. m = added mass,
K = radius of gyration of added mass.

$$\psi = \frac{k_2 S_2}{k_1 S_1} \sqrt{\frac{Y_2 \rho_2}{Y_1 \rho_1}} \quad ; \quad \chi = \sqrt[4]{\frac{\rho_2 Y_1 k_1^2}{\rho_1 Y_2 k_2^2}} \qquad .$$

change of cross-section

$$\tau = \left[\frac{\sigma^{-1,25} + \sigma^{-0,75} + \sigma^{0,75} + \sigma^{1,25}}{0,5\sigma^{-2} + \sigma^{-0,5} + 1 + \sigma^{0,5} + 0,5\sigma^2} \right]^2$$

change of material

$$\tau = \left[\frac{2\sqrt{\chi\psi}(1+\chi)\cdot(1+\chi)}{\chi(1+\psi)^2 + 2\psi(1+\chi^2)} \right]^2$$

corner

$$\tau = 2\left[\sigma^{-1,25} + \sigma^{1,25}\right]^{-2}$$

junction

$$\tau_{12} = \frac{1}{2}\left[\sigma^{-1,25} + \sigma^{1,25}\right]^{-2} \; ; \quad \tau_{13} = \frac{1}{2}\left[1 + 2\sigma^{5/2} + \sigma^5\right]^{-1}$$

added mass

$$\tau \approx 1; \text{ for } f < 0,5\,fs; \quad fs = \frac{1}{2\pi}\frac{K_1}{K_2}\sqrt{\frac{Y_1}{\sigma_1}}$$

$$\tau \approx \left[1 + \frac{f}{f_u}\right]^{-1}; \text{ for } f > 2fs; \quad f_u \approx \frac{2\rho_1 S_1^2 K_1 \sqrt{Y_1\rho_1}}{\pi^2 m}$$

If numbers are put into the formulas it can be seen, that
changes in cross-section or material and corners usually
give only a rather small vibration reduction. Added masses
have a small effect at low frequencies but at higher frequencies when the added mass is larger than the mass of the beam
over half a wavelength a good vibration reduction can be
obtained, provided that the mass is rigidly connected to
the beam.

When the beams on both sides of the discontinuity are not
very long or highly damped, resonance effects occur, which
further decrease the vibration reduction. This explains
why a periodic system (ship-hull etc.) does not have the
combined vibration reduction of many discontinuities.
Another effect which partly explains the strong transmission of structure borne sound over large distances is
the excitation of other wave-types especially longitudinal
waves. Thus it is not much of a surprise that in large,
undamped structures such as ship hulls, airplane fuselages
or buildings the structure borne sound usually decays at
a rate of less than one dB/m.

10. Radiation of structure borne sound

10.1 The basic mechanism

The problem of sound radiation from a vibrating surface
/1/ is a difficult one because the radiated power depends
on the mean square velocity of the vibrating surface as
well as on the details of the spatial distribution of the

velocity. This can be seen immediately, when we remind
ourselves, that for sound energy to be generated compression
of air is necessary (see Fig. 15).

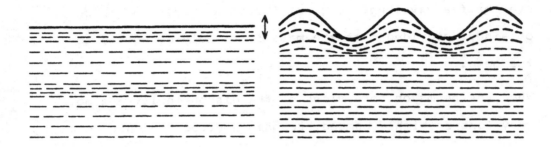

Fig. 15: Alternating pressure in front of a vibrating
         surface
         left: The air is compressed (sound radiation)
         right: The air moves between regions of opposite
                phase (very small sound radiation)

In order to take into account this effect a quantity called
radiation efficiency is introduced. It is defined as

$$\sigma = \frac{P}{\rho_0 c_0 \, v^2 S} \quad . \qquad (67)$$

Here P is the radiated power, S is the area of the radiating
surface v^2 is its mean square velocity. If there is no
hydrodynamic short-circuit (left side of Fig. 15) one would
expect $\sigma = 1$ because when the surface vibrates the air
must be compressed. If there is, however, a hydrodynamic
short-circuit the radiation efficiency is less than one.
There are just a few rare cases where $\sigma > 1$.

For noise control purposes it is desirable to make σ as small as possible; i.e. to have for each vibrating element a close neighbour that moves in opposite phase. For rigid bodies this is achieved by making them smaller than a quarter of a wavelength, for vibrating surfaces it is necessary that the surface vibration has a wavelength which is less than the wavelength of sound in air. Examples of measured radiation efficiencies see Fig. 16 and 17.

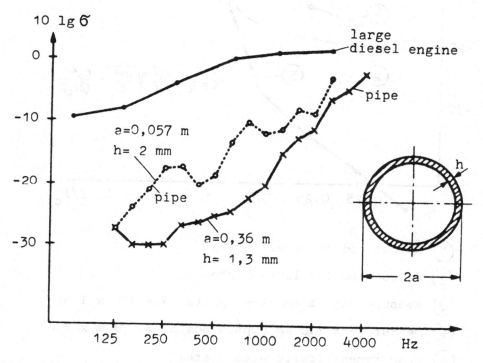

Fig. 16: Examples of radiation efficiencies

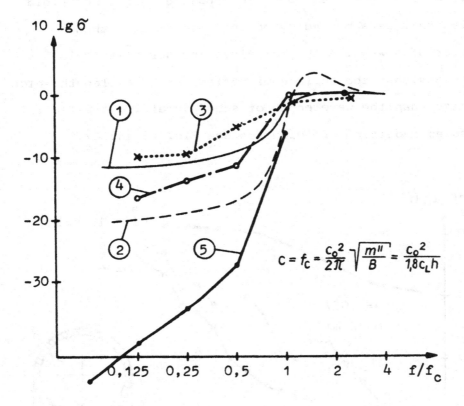

$$C = f_c = \frac{c_0^2}{2\pi} \sqrt{\frac{m''}{B}} = \frac{c_0^2}{1{,}8c_L h}$$

① calculated for small plate

② calculated for large plate

③ measurement (5 mm steel plate, S = 1 m x 1 m)

④ measurement (10 mm gypsum board, S = 3 m x 3 m)

⑤ measurement (perforated plate)

Fig. 17: Radiation efficiencies of plates

10.2 Examples of radiation efficiencies

A small rigid body (sphere) that vibrates back and forth (dipole) has a radiation efficiency given by

$$\sigma = \frac{\omega^4 a^4}{12\, c_o^4} \quad \text{for } a < \lambda/6 \qquad (68)$$

$$\sigma = 1 \quad \text{for } a > \lambda/2 \quad .$$

For a vibrating cylinder (pipe, wire) the corresponding formulas are

$$\sigma = \frac{\pi}{2}\ \frac{\omega^3 a^3}{c_o^3} \quad \text{for } a < \lambda/6 \qquad (69)$$

$$\sigma = 1 \quad \text{for } a > \lambda/2 \quad .$$

Here a is the radius. Cylinders and spheres with higher order motion ($2\,n$ nodal lines around a circumference) have a very small radiation efficiency if $a < n\lambda/6$. When the diameter is large or the frequency high ($a > n\lambda/2$) the radiation efficiency is around unity.

For plates in bending motion σ is determined by the ratio of the bending wavelength λ_B to the wavelength λ in the surrounding air. When $\lambda_B > \lambda$, the radiation efficiency is one or slightly above one. Since

$$\lambda_B = \frac{1}{2\pi} \sqrt[4]{\frac{B}{\omega^2\, m''}} \qquad (70)$$

(B = bending stiffness, m" = mass per unit area) it is found that bending waves are radiated strongly if

$$f > f_c = \frac{c_o^2}{2\pi} \sqrt{\frac{m''}{B}} \quad . \qquad (71)$$

The frequency f_c occurring here is identical with the critical frequency that is important for the transmission loss of single walls. Eq. (71) shows, that the critical frequency is high when the plate is thin (or limp); thus for a given velocity thin or limp plates radiate less sound than thick or stiff ones.

In the frequency range $f < f_c$ an infinitely large homogeneous plate would not radiate any sound at all because the hydrodynamic short-circuit would be perfect. Finite plates, however, and especially plates with discontinuities such as stringers, stiffeners etc. do radiate some sound because in some regions the short-circuit is not complete.

A reasonably good formula for estimating σ in the range $f < f_c$ is

$$\sigma \approx \frac{U c_o}{\pi^2 \, S \, f_c} \sqrt{\frac{f}{f_c}} \quad . \qquad (72)$$

Here U is twice the length of the stiffeners plus the length of the boundary and S the plate area.

Measured values of σ are shown in Fig. 17. In the light of the above discussion it is not surprising, that a perforated plate - with almost ideal short-circuit between the front and the back of the plate has a very small radiation efficiency.

10.3 Special cases

The formulas given in the last section hold only when the velocity field on the vibrating surface is fairly uniform. For lightly damped constructions this is usually the case. If however, the velocity changes considerably (more than 10 dB) over the vibrating surface the concept of a mean square velocity almost loses its meaning and it cannot be expected that eq. (67) is still useful. In fact even the proportionality between P and $\overline{v}^2$ may be lost, simply because the regions which determine $\overline{v}^2$ are not those from where the sound is radiated. A typical example of that is a large, highly damped plate which is excited at one point; obiously $\overline{v}^2$ is determined by the whole plate area but the sound radiation stems only from the region near the point of excitation. It is therefore not a surprise that for large, highly damped plates below the critical frequency there is a lower limit for the radiated sound power if the excitation is by a point force F. This lower limit is given by

$$P_F = \frac{F^2 \rho_o}{2 \pi c_o m''^2} \quad . \qquad (73)$$

This fact explains why additional damping reduces the mean square velocity of a vibrating structure but sometimes may not reduce the radiated sound because the force and consequently the limiting power P_F are not affected by damping.

Another example where formula (72) may not be true is when the excitation is by air borne sound; in this case there are forced waves (with a long wavelength) and free bending

waves (with a shorter wavelength) simultaneously on the
plate and consequently eq. (72) gives values which are too
small.

References

/1/ Cremer, L., Heckl, M., Ungar, E.: Structure-Borne-Sound

                                  Springer 1973

/2/ Leissa, A.W.: Vibration of Plates.

                                  NASA SP - 160 (1969)

/3/ Leissa, A.W.: Vibration of Shells.

                                  NASA SP - 288 (1973)

/4/ Lyon, R.H.: Statistical Energy Analysis of Dynamical

         Systems.

         MIT-Press (1975)

/5/ Heckl, M., Nutsch, J.: Körperschalldämmung und

         -dämpfung.

         Kap. 21 in Taschenbuch der Technischen

         Akustik (ed. Heckl, Müller). Springer 1975

/6/ Ross, D., Ungar, E., Kerwin, E.: Damping of Plate

         flexural Vibrations by means of

         Viscoelastic lamina.

         Appeared in Structural Damping (ed. Ruzicka)

         Amer. Soc. Mechn. Eng. (1959)

/7/ Ross, D.: Mechanics of Underwater Noise.

         Pergamon Press 1976

Transmission loss of walls and enclosures

M. Heckl, Technische Universität
Berlin

The main purpose of partitions is to reflect incoming sound
so that it cannot propagate into regions where it is unwanted. The sound energy remains almost unchanged.
($P_i = P_r + P_t$)

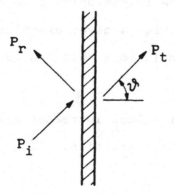

Fig. 1: Sound transmission through a plane wall

The sound insulation properties of partitions are characterised by the transmission coefficient

$$\tau = P_t/P_i \qquad (1)$$

or by the transmission loss

$$TL = 10 \log 1/\tau = 10 \cdot \lg(P_i/P_t), \quad (2)$$

Since the transmission loss changes considerably with frequency the values of TL usually are given as a frequency
curve. For rough estimates and quick comparisons it can be
sufficient to use an average value of TL called $\overline{TL}$, which is
obtained by taking the arithmetic mean of the values of TL
in the frequency range from 125 Hz to 4000 Hz.

It should be noted, that the transmission loss is not identical with the level difference on both sides of the partition and may deviate considerably from the insertion loss
(see part 5 on enclosures).

1. Transmission loss of single walls

Fig. 2 and 3 show the transmission loss of walls made of
building materials, Fig. 4 gives examples for thin steel
plates. The main conclusions that can be drawn out of the
figures are /1, 2/

a) the transmission loss increases with the mass per
unit area of the partition.

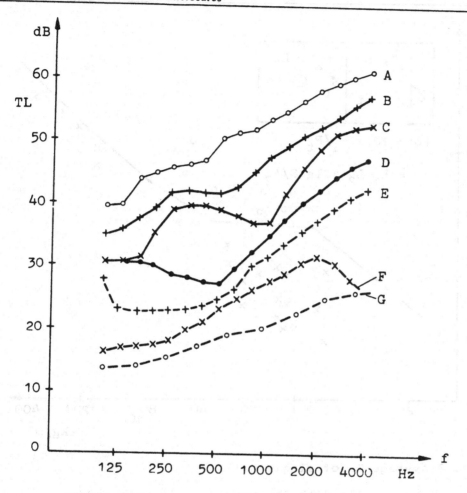

Fig. 2 TL of homogeneous, single walls

A: 250 mm brick, 480 kg/m^2, $\overline{TL}$ = 53 dB

B: 125 mm brick, 260 kg/m^2, $\overline{TL}$ = 46 dB

C: 85 mm brick, 170 kg/m^2, $\overline{TL}$ = 40 dB

D: 40 mm concrete, 95 kg/m^2, $\overline{TL}$ = 35 dB

E: 100 mm gypsum, 80 kg/m^2, $\overline{TL}$ = 29 dB

F: 8 mm gypsum, 7 kg/m^2, $\overline{TL}$ = 25 dB

G: 3 mm cardboard, 4 kg/m^2, $\overline{TL}$ = 20 dB

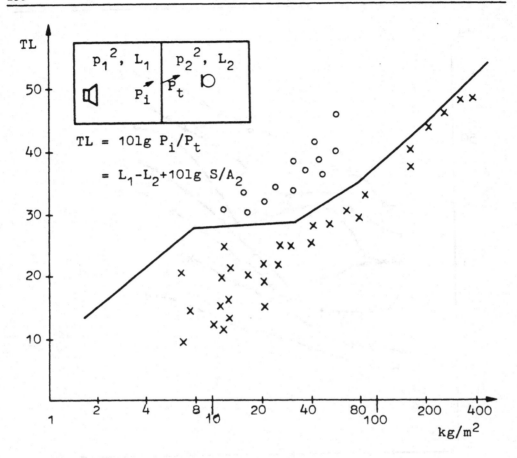

Fig. 3 Average TL of single walls

    o o o   walls with low bending stiffness and/or
            high damping

    x x x   stiff walls

    ───────   usual building materials

L_1, L_2 = sound pressure levels, S = wall area,
A_2 = absorption area

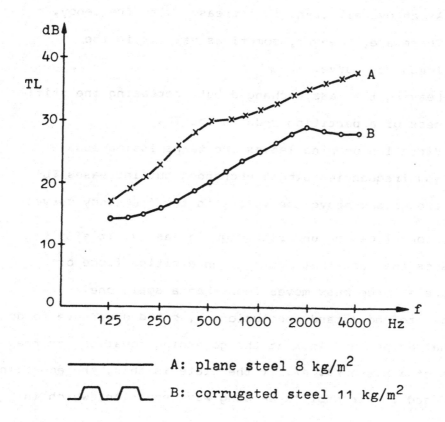

A: plane steel 8 kg/m²
B: corrugated steel 11 kg/m²

Fig. 4 Comparison of the transmission loss of a
       limp and a stiffened steel plate

b) As an overall trend TL increases with frequency.
   There are, however, sometimes valleys in the
   frequency curve.

c) Leaving the mass unchanged but increasing the stiffness of a partition reduces the TL.

d) Vibration damping leaves the transmission loss at
   low frequencies almost unchanged but increases TL
   around and above the valley in the frequency curve.

Conclusion a) can be understood quite easily, it simply
expresses the fact that for a given exciting force or
pressure a large mass moves less than a small one.
In order to understand conclusions b, c and d we have to go
somewhat deeper and look at the governing equation for the
motion of a plate /3, 4/. If the plate is thin, the equation
to be used is Kirchhoff's bending wave equation, which in
the one dimensional case reads

$$B \frac{\partial^4 \xi(x)}{\partial x^4} + m' \frac{\partial^2 \xi(x)}{\partial t^2} = p(x), \tag{3}$$

Here B is the bending stiffness and m' the mass per unit
area of the plate, $\xi(x)$ is the plate motion and p(x) the
exciting pressure, which is the difference of the sound
pressures on both sides of the plate. For purely sinosoidal
motion eq. (3) becomes

$$B \frac{d^4 \xi(x)}{dx^4} - \omega^2 m' \xi(x) = p(x). \tag{4}$$

For the simplest case of a plane wave coming under an
angle ϑ the exciting pressure distribution is

$$p(x) = p_o \, e^{-j\frac{\omega}{c_o} x \sin \vartheta} \; .$$

Assuming the same spatial distribution for the displacement
we find

$$\xi = \frac{p_o}{B\omega^4 \sin^4 \vartheta / c_o^4 - \omega^2 m'} \qquad (5)$$

This equation shows, that there exist frequencies for which
the denominator can vanish, i.e. for which the plate motion
can become rather big. These frequencies are called coincidence frequencies; they are given by

$$\omega_\vartheta = \sqrt{\frac{m'}{B}} \; \frac{c_o^2}{\sin^2 \vartheta} \; . \qquad (6)$$

Since $\sin \vartheta$ is always less than unity this coincidence can
never occur below a certain limiting value, which is called
the critical frequency

$$\omega_c = \sqrt{\frac{m'}{B}} \, c_o^2 \; . \qquad (7)$$

For $\omega < \omega_c$ eq. (5) can be approximated by $\xi_o = p_o / \omega^2 m'$;
i.e. the motion is purely mass controlled. For $\omega > \omega_c$
however there is always an angle for which ξ_o becomes very
large giving rise to a small TL. This effect is especially
pronounced in the vicinity of ω_c, where the TL is always
low even when sound comes in from all directions.

For air-borne sound transmission and for homogeneous walls
the critical frequency can be found from fig. 5.

For nonisotropic walls (corrugated plates) there are at least
two critical frequencies which can be obtained from eq. (7)
by introducing the bending stiffness in the different directions. If the internal damping - characterized by the
loss factor η - of the plate is high the coincidence effect
is less pronounced because in this case B has to be replaced
by B $(1 + j \eta$) and absolute values have to be taken. Thus
we get

$$|\xi_o|^2 = \frac{p_o^2}{(B\omega^4 \sin^4 \vartheta /c_o^4 - \omega^2 m')^2 + \eta^2 B^2 \omega^8 \sin^8 \vartheta /c_o^8} \cdot (8)$$

The denominator never can vanish and therefore the coincidence effect is the less pronounced the higher the loss
factor of the plate material. /5/

2. Double walls

The transmission loss of double walls as compared to that
of single walls is shown in fig. 6. It can be seen that
in a certain frequency range the TL of a double wall is
better, in another range it is worse than that of a simple
wall /6/. Therefore it is important to known where the
frequencies of poor TL are. Since the low TL is caused
by a resonance of the masses of the walls and the stiffness of the intermediate layer the region of low TL is
around the frequency

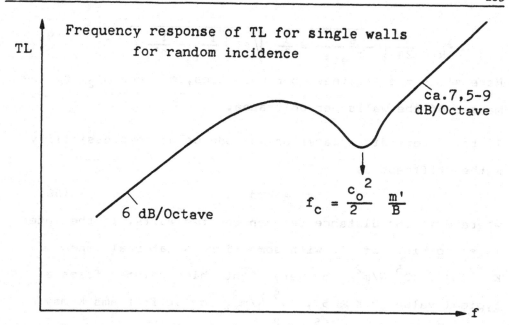

Frequency response of TL for single walls
for random incidence

ca.7,5-9
dB/Octave

6 dB/Octave

$$f_c = \frac{c_o^2}{2} \frac{m'}{B}$$

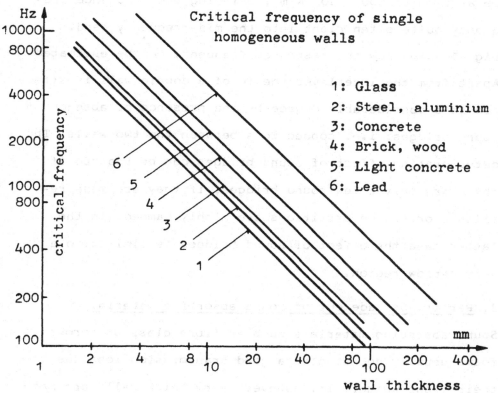

Critical frequency of single
homogeneous walls

1: Glass
2: Steel, aluminium
3: Concrete
4: Brick, wood
5: Light concrete
6: Lead

wall thickness

Fig. 5

$$f_R = \frac{1}{2\pi} \sqrt{\frac{s'}{m'_{eff}}} = \frac{1}{2\pi} \sqrt{s' \left(\frac{1}{m'_1} + \frac{1}{m'_2}\right)} \quad . \qquad (9)$$

Here s' is the stiffness per unit area, m'_1 and m'_2 are the masses of the walls per unit area.

If the intermediate layer has a modulus of compressibility K the stiffness is

$$s' = K/d \quad , \qquad (10)$$

where d is the distance between the two walls. If the inter-layer consists of air with some fibrous material, then $K = 1.4 \cdot 10^5$ N/m^2; for very light, high polymer foams a typical value is $K \approx 5 \cdot 10^5$ N/m^2, for stiff foams K may be as high as $500 \cdot 10^5$ N/m^2, bringing the resonance frequency quite often right into the mid-frequency band. Fig. 7 shows how the resonance frequency can be estimated. Apart from the interlayer the TL of a double wall is affected to a considerable degree by the presence or absence of sound bridges; i.e. connections between the two walls. The deteriorating effect of sound bridges can be reduced if there are only a few sound bridges, if they are made re-silient or if the partitions are highly damped. In the latter case the effect of sound bridges is limited to a very narrow region.

3. Partitions made out of sound absorbing material
Sound absorbing materials such as fibre glass or porous foams usually do not give a good transmission loss due to their light weight. If, however, very thick walls are made out of such materials the dissipation of sound inside

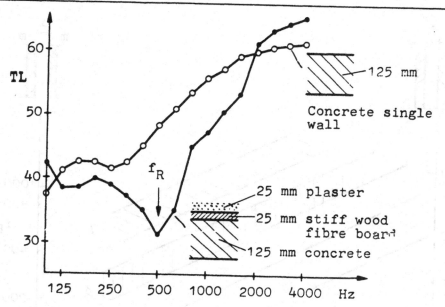

Fig.6a: Influence of a very stiff interlayer on the transmission loss of a concrete wall

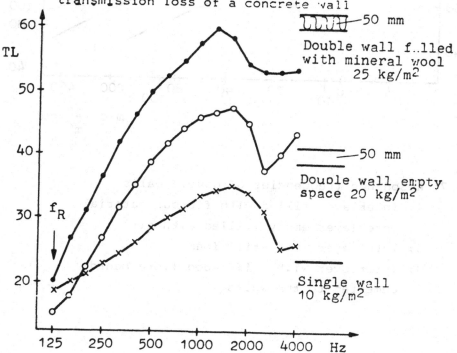

Fig. 6: Transmission loss of walls consisting of 125 mm sheet rock

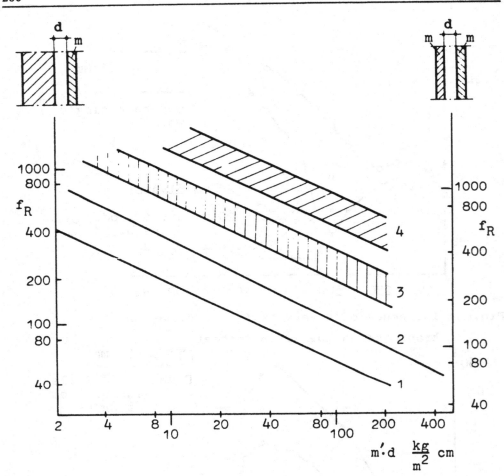

Fig. 7: Resonance frequencies of double walls
        1: Interlayer filled with fibrous material
        2: Interlayer empty (filled with air)
        3: Interlayer with stiff foam
        4: Interlayer with stiff wood fibre boards
           cemented to the walls

the wall can be so high that the sound level differences
between two rooms separated by such a thick wall can give
rise to rather large values. Since the decay of sound inside the wall material is determined by the thickness/wave-
length ratio, the transmission loss is especially high for
thick walls at high frequencies. Examples are shown in
fig. 8. /7/

As an application of such "walls" fig. 9 gives some results
for the sound transmission via a suspended ceiling. (E.g.
office rooms.) It is seen that a fibre glass "wall" above
the suspended ceiling considerably reduces the sound transmission. /8/

4. Pipes

The transmission loss of pipes of diameters up to 1 m is
different from the TL of plane walls, because pipes vibrate
in only a few well defined modes and therefore the elasticity plays a very important role especially at low frequencies. /9/

The TL of pipes from the inner side to the outer side is
shown in fig. 10 under the assumption that there is air
inside and outside the pipe. The most striking result is
that the TL is very low when $f \approx \sqrt{E/\rho}/2 \cdot \pi a$; that is the
frequency where one longitudinal wave length in the plate
material is just as long a the circumference. For steel
pipes of 25 cm diameter the frequency of low TL is around
3 kHz. It is therefore not surprising that in the vicinity

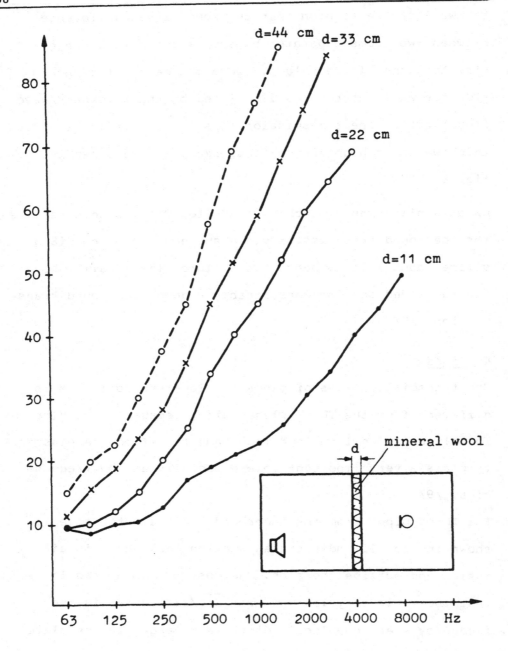

Fig. 8: Transmission loss of walls consisting of mineral wool with a density of 60 kg/ m^2

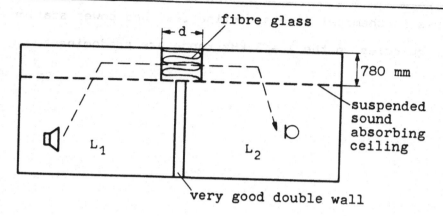

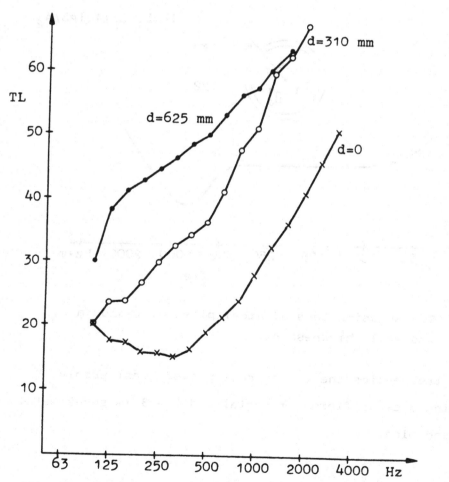

Fig. 9: Influence of a fibre glass barrier ("wall")
on the sound transmission via suspended ceilings

of pipes in chemical plants, refineries, and power stations
the frequencies in the 2 - 5 kHz range are predominant.

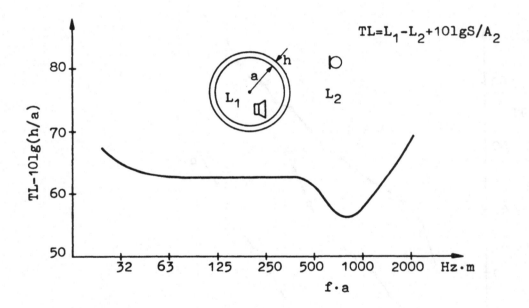

$$TL = L_1 - L_2 + 10 \lg S/A_2$$

Fig. 10: Transmission loss of steel pipes of radius a
and wall thickness h.

At high frequencies the TL can be improved considerably
by putting 5 cm of fibrous material and 1 - 3 mm sheet metal
around the pipe.

5. Enclosures

The purpose of enclosures is to reduce the noise at some distance from the source. To this end walls of high TL are necessary but not sufficient as can be seen from the following calculations.

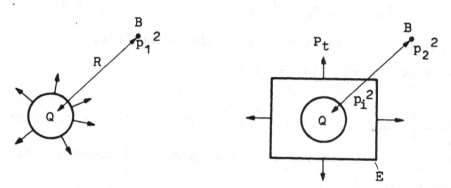

Source without enclosure   Source with enclosure

Fig. 11: Insertion loss by enclosures

Q = Source, B = receiver, E = enclosure

If we are in a free space (the conclusions hold as well for a bounded space) and if the omnidirectional sound power radiated from the source is P the sound pressure p_1 at the receiver at a distance R is

$$p_1^2 = \rho_o c_o P/4 \pi R^2 \qquad (11)$$

(ρ_o, c_o density and speed of sound in air). With the enclosure the sound pressure is reduced to

$$p_2^2 = \rho_o c_o P_t/4 \pi R^2 = \rho_o c_o \tau P_i/4 \pi R^2 . \qquad (12)$$

(See e.q. (1)). The incident sound power P_i on the inner side of the enclosure is related to the surface S of the enclosure and the sound pressure p_i inside the enclosure by

$$P_i = \frac{P_i^2}{\rho_o c_o} \frac{S}{4} \cdot \qquad (13)$$

Assuming (as is usually the case) that the power output of the source is not affected by the enclosure, P_i is given by

$$P_i^2 = \rho_o c_o P \frac{4}{A} , \qquad (14)$$

where A is the absorbing area on the inner side. Combination of eq. (11) - (14) gives

$$\frac{P_1^2}{P_2^2} = \frac{A}{\tau S} \text{ or IL} = 10 \lg P_1^2/P_2^2 = TL + 10 \lg A/S \cdot \quad (15)$$

Here IL is the insertion loss, i.e. the level reduction at the point of reception.

The main result of the little calculation is that the insertion loss IL depends on the TL of the enclosure walls and the average absorption coefficient A/S inside the enclosure. Therefore when enclosures with little absorption are used it is very easily possible that the insertion is well below the transmission loss.

Apart from the TL and the absorption coefficient another important aspect is the presence of slits which quite often are unavoidable. As a rule of thumb one may say that a slit or area S_s reduces the TL to

$$10 \lg S/S_s \cdot \qquad (16)$$

The deteriorating effect of slits can be reduced if they are made sound absorbing. An example is shown in the next figure.

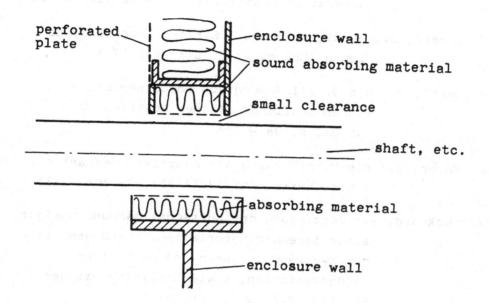

Fig. 12: Sound absorbing slit of an engine enclosure

Sound absorbing channels or other types of silencers are also necessary for ventilation of the enclosure (enclosures usually have good heat isolation properties).

References

/1/ Gösele, K.: Schall, Wärme, Feuchtigkeit. Grundlagen, Erfahrungen und praktische Hinweise für den Hochbau. Bauverlag Wiesbaden 1977

/2/ Ver, I.L., Holmer, C.I.: Interaction of Sound waves with solid structures. Chap. 12 in Noise and Vibration Control (ed. Beranek) McGraw Hill 1971

/3/   Cremer, L.: Theorie der Luftschalldämmung dünner
          Wände bei schrägem Einfall.
          Akustische Zeitschrift 7 (1942) S. 81-104

/4/   Cremer, L., Heckl, M., Ungar, E.E.: Structure Borne
          Sound. Chapt. VI,7 Springer 1973

/5/   Heckl, M.: Die Schalldämmung von homogenen Einfach-
          wänden endlicher Fläche. Acustica 10
          (1960), p. 98 - 108

/6/   Meyer, E.: Die Mehrfachwand als akustisch-mechanische
          Drosselkette. ENT 12 (1935), S. 303 - 333

/7/   Rückward, W.: Bestimmung der Luftschalldämmung poriger
          absorbierender Materialien, Messungen nach
          dem Zweiraumverfahren und nach einem
          Sondenverfahren, sowie Vergleich mit der
          Theorie. Private communication

/8/   Mechel, F.: L'isolation acoustique des plafonds
          suspendus. Private communication

/9/   Heckl, M.: Experimentelle Untersuchungen zur Schall-
          dämmung von Zylindern. Acustica 8 (1958)
          S. 259 - 265.

Noise reduction by sound absorbing materials

M. Heckl, Technische Universi-
tät Berlin

1. Basic mechanisms and empirical data

The safest way to reduce noise is to transfer the sound
energy into heat. Since the dissipation of sound when it
propagates through an unbounded medium like air or water
is extremely small, one has to rely almost exclusively
on viscous and thermal losses at the boundaries between
solids and the surrounding medium in order to absorb air
borne sound. The simplest case of the absorption mechanism
is shown in the next figure

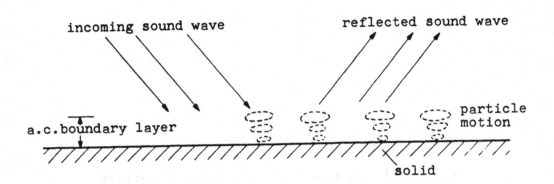

Fig. 1: Transfer of sound energy into heat at a
        solid-gaseous interface

When the motion of vibrating air particles has a component
parallel to the boundary, there is some friction and consequently some loss of sound energy in a very thin a.c. boundary layer near the solid. At the frequencies of interest
these losses are extremely small; therefore a noticeable
absorption is achieved only when the surface exposed to the
sound is increased considerably. This is done by the use of
fibrous [+] or other porous materials, which are contained
in all common sound absorbing devices. Quite often these
porous sound absorbing materials are not exposed directly
to the sound field, because they may be blown away or for

[+] A glass-wool mat of 5 cm thickness and 1 m^2 area consisting
    of fibres of $5 \cdot 10^{-6}$ m has fibres with a total length of
    more than 100 000 km and an overall fibre surface of more
    than 1000 m^2.

aesthetic reasons. In such cases sound absorbing materials
are covered by perforated plates, by foils or by thin plates.
This way the following types of sound absorbers are obtained (see also table of absorption coefficients). /1/, /2/

a) High frequency absorbers

Fibrous mats, foams with open pores, porous stones etc.
of several centimeter thickness - possibly with some
airspace behind them - are very effective high frequency absorbers. If they are covered by a thin cloth or by
a highly perforated (more than 20 % perforation) plate
with small holes (less than 3 mm diameter) the absorption properties remain almost unchanged.

b) Mid-frequency absorbers

If a fibrous mat is covered by a thin foil or by a plate
of less than 10 % perforation the absorption at the high
frequencies is reduced but in the mid-frequency region
it is increased. The frequency of maximum absorption is
given by

$$f_m = \frac{600}{\sqrt{md}} \quad (\text{d in cm, m in kg/m}^2). \qquad (1)$$

Here d is the distance between the foil or plate and the
rigid backing. When foils are used, m is their mass per
unit area; when perforated plates of thickness h, hole
radius a and perforation ratio ε are used the effective mass is

$$m \approx \frac{\rho_o}{\varepsilon} \, (h + 1.6 \, a). \qquad (2)$$

c) Low frequency absorbers

If sound absorbing material is covered by a thick foil

or by an unperforated plate, the absorption in the high-
and mid-frequency band is reduced considerably. The remaining absorption is centered around the frequency f_m
given by eq. 1.

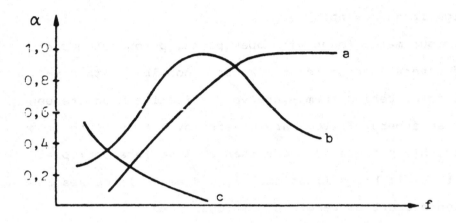

Fig. 2: Absorption coefficient of the three types of sound
absorbers

Examples of absorption coefficients /3/

	125	250	500	1000	2000	4000 Hz
Wallpaper on plaster	0,02	0,03	0,04	0,05	0,06	0,07
Mineral wool 30 kg/m³, 30 mm thick, no air-space 10,6 Rayl/cm	0,16	0,25	0,6	0,75	0,8	0,9
Mineral wool 100 kg/m³, 50 mm thick, no air space 22 Rayl/cm	0,25	0,6	1,0	1,0	1,0	0,97
Mineral wool 46 kg/m³, 30 mm thick, 50 mm air-space, 12 Rayl/cm	?	0,75	1,0	0,9	0,82	0,84
Mineral wool, covered by a perforated plate 15 mm hole-diameter, 18 % perforation	0,30	0,69	1,0	0,8	0,66	0,62
Mineral wool (5 cm thick) covered by a foil of 0,2 kg/m²	0,3	0,7	1,0	0,7	0,5	0,2
Concrete floor with 3 - 4 mm carpet	0,02	0,03	0,05	0,13	0,35	0,5
Concrete floor with 8 - 10 mm carpet	0,03	0,08	0,2	0,35	0,65	0,9
Plywood plate of 6 mm thickness, backed by 50 mm mineral wool	0,57	0,37	0,15	0,07	0,05	0,03

2. Noise reduction by sound absorption

If sound absorbing material, e.g. a sound absorbing ceiling
is brought into a room two effects are observed /1, 2, 4/

a) the reverberation time is reduced

b) the noise levels go down somewhat.

In rooms of moderate dimensions (office, small factory) the
change in reverberation time is given by

$$\frac{1}{T_2} = \frac{1}{T_1} + \frac{C,16\ V}{A} \tag{3}$$

and the change in level by

$$\Delta L = 10 \lg \frac{A + \Delta A}{A} = 10 \lg \frac{T_1}{T_2} . \tag{4}$$

(V in m^3, A, ΔA in m^2, T in sec)

Here $\Delta A = \Sigma \alpha_i S_i$ is the absorption area that has been
brought into the room, which originally had an absorption
area A. V is the room volume, T_1 and T_2 are the reverberation times before and after bringing in the additional absorption.

Usually it is difficult to make ΔA bigger than five times
the original absorption area A, therefore the level reduction
by an absorbing ceiling or the like seldom is larger than
8 dB and most often only around 3 dB. It should be noted
also that in the immediate vicinity of the sound source
(i.e. for distance smaller than $0,15 \cdot \sqrt{A + \Delta A}$) there is
practically no sound reduction, because the direct sound
field is dominating.

In long rooms with small to moderate height (production
lines, big offices etc.) the effect of a sound absorbing
ceiling can be much higher especially when many scattering
objects such as machines, furniture, sound shields etc.
are in the sound path.

In such cases the room behaves almost like a sound absorbing duct, giving a rather strong decay of sound level
with increasing distance from the source. See fig. 3

3. Dissipative mufflers

The purpose of silencers is to prevent the transmission of
sound through ducts without seriously obstructing the flow of
air or gas through the duct. This objective can be achieved
by reflecting the sound or by transferring sound energy
into heat (dissipation). Here we shall restrict ourselves
to dissipative mufflers. These mufflers which are used
very much in ventilation systems, in intake our exhaust
ducts of power station, gasturbines etc., consist mainly
of a very effective sound absorbing lining that is exposed
to the sound as much as possible. [+)]

[+)] Practical problems of silencer construction, the difficulties associated with dust particles that tend to close
the pores of sound absorbing materials and the effect
of flow on the silencer performance are not considered
here

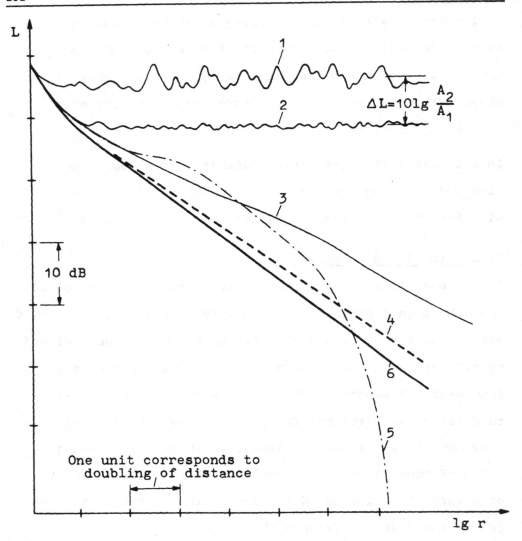

Fig. 3 Decay of sound levels in rooms

1: small room, little absorption
2: small room, strong absorption
3: flat room, little absorption on ceiling ($\alpha \approx 0,1$),
   strong absorption on walls, no scatterers
4: flat room, strong absorption on ceiling ($\alpha > 0,8$),
   no scatterers
5: flat room, strong absorption, many scatterers
6: free space propagation

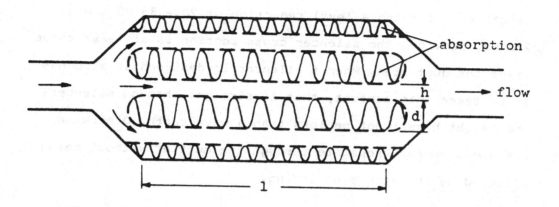

Fig. 4 Dissipative muffler

As a rule of thumb it can be said, that the level reduction
by a muffler is given by

$$\Delta L \approx 3\, \alpha\, \frac{l}{h} = 1,5\, \alpha\, \frac{Ul}{S} \, . \qquad (5)$$

The first expression holds for rectangular ducts, the second
one for other geometries. U is the circumference of the absorbing lining, S the area through which the gas flows,
α is a quantity less than unity; it is related to the absorption coefficient of the lining. Eq. (5) shows that a
big noise reduction is achieved when the muffler is long
(in terms of the duct height h)and when an effective lining
is used; since a good, broad band absorption at low frequencies can be obtained by thick linings only, mufflers
which are effective at low frequencies must have thick
linings (e.q. $d > \lambda/6$). In practice the duct height has to
be above a certain limit in order to keep the pressure
drop low; therefore silencers often are of considerable

length. Silencers in ventilation ducts typically have a

length of 1-2 m and a level reduction of 20 - 35 dB above

250 Hz, provided the silencer cross section is at least three

times the duct area. When a silencer is designed it also has

to be taken into account, that in straight-through silencers

the height h must not be less than a wavelength. Otherwise

a sound beam can propagate through the muffler without being

affected by the absorbing lining.

For a more detailed theory /5, 6/ of dissipative mufflers

it is necessary to go back to the wave equation for harmonic

waves of angular frequency ω in a rectangular duct. When c_o

is the speed of sound this equation for the sound pressure

reads

$$\frac{\partial^2 p}{\partial x^2} + \frac{\partial^2 p}{\partial y^2} + \frac{\omega^2}{c_o^2}\, p = 0 \qquad\qquad (6)$$

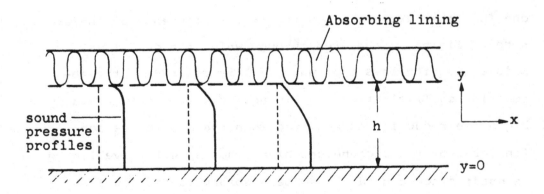

Fig. 5: Wave propagation in a lined duct

Assuming that a sound wave propagates in positive

x-direction a possible solution of eq. (6) is:

$$p = p_o\, e^{-\gamma x} \cos(\varepsilon y), \qquad\qquad (7)$$

provided that

$$\varepsilon^2 = \gamma^2 + \omega^2/c_o^2 \, . \tag{8}$$

Here we are interested in the quantity γ because its real part determines the decay of sound as x is increased.

In order to find ε we calculate the sound pressure and the velocity at the line $y = h$:

$$p \, (y = h) = p_o e^{-\gamma x} \cos (\varepsilon h) \tag{9}$$

$$v \, (y = h) = \frac{-1}{j\omega\rho} \frac{\partial p(y = h)}{\partial y} = \frac{1}{j\omega\rho} p_o e^{-\gamma x} \cdot \varepsilon \sin(\varepsilon h) \, .$$

The ratio $p(y=h)/v(y=h)$ must be equal to the acoustic impedance of the absorbing lining. This impedance is a complex quantity which depends on the thickness of the lining, the material, etc.. Eq. (9) and the relation

$$\frac{p(y = h)}{v(y = h)} = Z \tag{10}$$

are combined to

$$\varepsilon h \cdot \tan (\varepsilon h) = j \, \omega \rho_o h/Z = j \, \frac{\omega h}{c_o} \, \frac{\rho_o c_o}{Z} \, . \tag{10}$$

This relation is somewhat difficult to solve for ε (which also is complex). A reasonable approximation can be obtained however by writing

$$\tan (\varepsilon h) \approx 0,1875 \, (\varepsilon h) - \frac{1,0047}{\varepsilon h + \frac{\pi}{2}} - \frac{1,0047}{\varepsilon h - \frac{\pi}{2}} \, ,$$

which gives

$$\varepsilon h = \frac{2,47 + \gamma \pm \sqrt{(2,47 + \gamma)^2 - 1,87 \, \gamma}}{0,38} \tag{11}$$

here $Y = j\frac{\omega h}{c_o} \cdot \frac{\rho_o\, c_o}{Z}$. Eq. (11) gives two complex values for
ε h; of practical importance is only that one, which gives
the lower attenuation.

Thus when the impedance of the lining is known the decay
rate Re $\{\gamma\}$ can be calculated. The calculations show that
there is an impedance for which the decay of sound in a duct
has a maximum (19 1/h dB). Unfortunately this optimum impedance can be realised for very narrow frequency bands only.
This is the reason, why sometimes different linings in one
silencer are used. If fibrous mats are used the optimum
impedance is obtained when the material is lightly packed.
Thus sometimes silencers can be improved by taking out some
sound absorbing material.

References

/1/ Cremer, L.; Müller, H.A.: Die wissenschaftlichen
              Grundlagen der Raumakustik Bd. I and II
              S. Hirzel, Stuttgart 1976

/2/ Beranek, L.L. (ed.): Noise and Vibration Control
              McGraw Hill 1971

/3/ Deutscher Normenausschuß: Schallabsorptionsgradtabelle
              Beuth Vertrieb, Berlin 1968

/4/ Kuttruff, H.: Room Acoustics
              Applied Science Publishers, London 1973

/5/ Cremer, L.: Theorie der Luftschalldämpfung im Rechteck-

              kanal mit schluckender Wand und das sich

              dabei ergebende Dämpfungsmaß.

              Acustica 3 (1953) p. 139 - 150

/6/ Mechel, F.: Schallabsorption, Schalldämpfer

              Kap. 18 u. 19 in Taschenbuch der

              Technischen Akustik (ed. Heckl, Müller)

              Springer, Berlin 1975

276 — G. BIANCHI: Noise Generation and Control
in Mechanical Engineering

ERRATA

This edition not for sale: numbering of some pages does not
respect actual sequence.

i) pages 117 - 120 should follow page 139
ii) page 221 should follow page 215

Printed in the United States
By Bookmasters